Selected Titles in This Series

(*[redacted]n the back of this publication*)

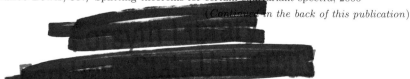

[redacted]

MEMOIRS
of the
American Mathematical Society

Number 715

Multi-Interval Linear Ordinary Boundary Value Problems and Complex Symplectic Algebra

W. N. Everitt
L. Markus

May 2001 • Volume 151 • Number 715 (first of 5 numbers) • ISSN 0065-9266

American Mathematical Society
Providence, Rhode Island

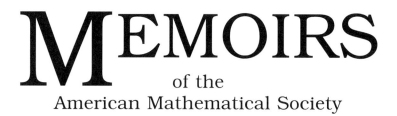

2000 *Mathematics Subject Classification.*
Primary 34B10, 51A50; Secondary 34L05.

Library of Congress Cataloging-in-Publication Data

Everitt, W. N. (William Norrie), 1924–
Multi-interval linear ordinary boundary value problems and complex symplectic algebra / W. N. Everitt, L. Markus.
p. cm. — (Memoirs of the American Mathematical Society, ISSN 0065-9266 ; no. 715)
"Volume 151, Number 715 (first of 5 numbers)."
Includes bibliographical references.
ISBN 0-8218-2669-7 (alk. paper)
1. Boundary value problems. 2. Differential operators. 3. Symplectic manifolds. I. Markus, L. (Lawrence), 1922– II. Title. III. Series.

QA3.A57 no. 715
[QA379]
510 s—dc21
[515′.35] 2001018141

Memoirs of the American Mathematical Society

This journal is devoted entirely to research in pure and applied mathematics.

Subscription information. The 2001 subscription begins with volume 149 and consists of six mailings, each containing one or more numbers. Subscription prices for 2001 are \$494 list, \$395 institutional member. A late charge of 10% of the subscription price will be imposed on orders received from nonmembers after January 1 of the subscription year. Subscribers outside the United States and India must pay a postage surcharge of \$31; subscribers in India must pay a postage surcharge of \$43. Expedited delivery to destinations in North America \$35; elsewhere \$130. Each number may be ordered separately; *please specify number* when ordering an individual number. For prices and titles of recently released numbers, see the New Publications sections of the *Notices of the American Mathematical Society.*

Back number information. For back issues see the *AMS Catalog of Publications.*

Subscriptions and orders should be addressed to the American Mathematical Society, P. O. Box 845904, Boston, MA 02284-5904. *All orders must be accompanied by payment.* Other correspondence should be addressed to Box 6248, Providence, RI 02940-6248.

Memoirs of the American Mathematical Society is published bimonthly (each volume consisting usually of more than one number) by the American Mathematical Society at 201 Charles Street, Providence, RI 02904-2294. Periodicals postage paid at Providence, RI. Postmaster: Send address changes to Memoirs, American Mathematical Society, P. O. Box 6248, Providence, RI 02940-6248.

This publication is indexed in *Science Citation Index*®, *SciSearch*®, *Research Alert*®, *CompuMath Citation Index*®, *Current Contents*®/*Physical, Chemical & Earth Sciences.*
Printed in the United States of America.

∞ The paper used in this book is acid-free and falls within the guidelines established to ensure permanence and durability.
Visit the AMS home page at URL: `http://www.ams.org/`

10 9 8 7 6 5 4 3 2 1   06 05 04 03 02 01

Dedicated to the memory of
I.M. Glazman, M.G. Krein and M.A. Naimark

Contents

Abstract

A multi-interval quasi-differential system $\{I_r, M_r, w_r : r \in \Omega\}$ consists of a collection of real intervals, $\{I_r\}$, as indexed by a finite, or possibly infinite index set Ω (where $\text{card}(\Omega) \geq \aleph_0$ is permissible), on which are assigned ordinary or quasi-differential expressions M_r generating unbounded operators in the Hilbert function spaces $L_r^2 \equiv L^2(I_r; w_r)$, where w_r are given, non-negative weight functions. For each fixed $r \in \Omega$ assume that M_r is Lagrange symmetric (formally self-adjoint) on I_r and hence specifies minimal and maximal closed operators $T_{0,r}$ and $T_{1,r}$, respectively, in L_r^2. However the theory does not require that the corresponding deficiency indices d_r^- and d_r^+ of $T_{0,r}$ are equal (*e. g.* the symplectic excess $Ex_r = d_r^+ - d_r^- \neq 0$), in which case there will not exist any self-adjoint extensions of $T_{0,r}$ in L_r^2.

In this paper a system Hilbert space $\mathbf{H} := \sum_{r \in \Omega} \oplus L_r^2$ is defined (even for non-countable Ω) with corresponding minimal and maximal system operators $\mathbf{T}_0$ and $\mathbf{T}_1$ in $\mathbf{H}$. Then the system deficiency indices $\mathbf{d}^{\pm} = \sum_{r \in \Omega} d_r^{\pm}$ are equal (system symplectic excess $Ex = 0$), if and only if there exist self-adjoint extensions $\mathbf{T}$ of $\mathbf{T}_0$ in $\mathbf{H}$. The existence is shown of a natural bijective correspondence between the set of all such self-adjoint extensions $\mathbf{T}$ of $\mathbf{T}_0$, and the set of all complete Lagrangian subspaces L of the system boundary complex symplectic space $\mathsf{S} = \mathbf{D}(\mathbf{T}_1)/\mathbf{D}(\mathbf{T}_0)$. This result generalizes the earlier symplectic version of the celebrated GKN-Theorem for single interval systems to multi-interval systems.

Examples of such complete Lagrangians, for both finite and infinite dimensional complex symplectic spaces S, illuminate new phenomena for the boundary value problems of multi-interval systems. These concepts have applications to many-particle systems of quantum mechanics, and to other physical problems.

Received by the editor 12 August 1999.

2000 *Mathematics Subject Classification.* Primary 34B10, 51A50; Secondary 34L05, 70H15.

Key words and phrases. Quasi-differential expressions, differential operators, symplectic algebras.

Multi-interval linear ordinary boundary value problems and complex symplectic algebra

1. Introduction: Goals, Organization

In this Introduction we review the role played by the Glazman–Krein-Naimark (GKN) Theorem in relating the self-adjoint operators generated in the boundary value theory for quasi-differential systems, defined on single intervals of the real line $\mathbb{R}$, to symplectic spaces; similarly for the generalizations to quasi-differential systems for multi-interval problems, again defined on $\mathbb{R}$.

A new kind of collective boundary value problems for ordinary differential expressions (formal differential operators), or more general quasi-differential expressions, on a countable set of intervals of $\mathbb{R}$, has been formulated by Everitt and Zettl, see [**14**] and especially [**13**] for examples involving an infinite number of intervals, and then resolved through their extension of the GKN-Theorem, see [**15**]. Such multi-interval problems arise in quantum mechanics of many particle systems, see [**2**], [**16**], [**17**] and [**18**], where boundary conditions refer to interfacial connections that link the intervals determined by the particles situated along a line; also in certain problems in applied mathematics, see for example [**3**], [**19**], [**20**] and [**21**].

Following Everitt-Zettl we shall here define general multi-interval quasi-differential systems

$$\{\, I_r, M_r, w_r : r \in \Omega \,\} \tag{1.1}$$

where Ω is a general but non-empty index set that may be finite, denumerable or non-denumerable; we often abbreviate (1.1) as a "multi-interval system". Such a multi-interval system (see precise details in Definition 2.1 below) consists of a set of prescribed intervals $I_r \subset \mathbb{R}$, each bearing a given positive weight w_r, so as to define the usual Hilbert function space $L^2_r(I_r; w_r) \equiv L^2_r$ of complex-valued, w_r-square-integrable functions on I_r, and each supporting an assigned quasi-differential expression M_r which thus generates appropriate (unbounded) linear operators in L^2_r—for each r in the prescribed index set Ω. It will be shown, under suitable hypotheses, that a multi-interval system (1.1) generates maximal and minimal operators $\mathbf{T}_1$ and $\mathbf{T}_0$, with domains $\mathbf{D}(\mathbf{T}_1)$ and $\mathbf{D}(\mathbf{T}_0)$ respectively, in the direct sum Hilbert space $\mathbf{H} = \sum_{r\in\Omega} \oplus L^2_r$; furthermore (1.1) generates self-adjoint operators in $\mathbf{H}$ which are determined by new kinds of generalized self-adjoint boundary conditions; for details see [**15**].

In previous papers, see [**7**],[**8**] and [**10**], and in the book [**9**] the present authors have recast the prior GKN-theory for the case of a single interval ($\Omega = \{1\}$) in terms of the geometry and linear algebra of finite dimensional complex symplectic spaces. In this paper we follow these now established notations, and use this same approach to develop our analysis of general multi-interval systems (1.1). In particular, the appropriate formulation of the GKN-Theorem is introduced using the concepts of infinite dimensional complex symplectic spaces, for example

$$\mathsf{S} = \mathbf{D}(\mathbf{T}_1)/\mathbf{D}(\mathbf{T}_0), \tag{1.2}$$

and their Lagrangian subspaces. Through these algebraic structures all the various kinds of self-adjoint boundary conditions can be described and classified for general multi-interval quasi-differential systems $\{I_r, M_{A_r}, w_r : r \in \Omega\}$, as described below in Definition 2.1. In this manner the goal of this current investigation is accomplished in Sections 5 and 6 below.

In Section 2 we give some definitions, and in Section 3 give a brief account of the required properties of complex symplectic spaces; further (also in Section 6) we introduce the topological and algebraic invariants for infinite dimensional complex symplectic spaces and their complete Lagrangian subspaces. In Section 4 we discuss the single interval quasi-differential system, in respect of symplectic spaces defined by (1.2), and in Section 5 extend these ideas to multi-interval systems (1.1). In the Section 6 we re-interpret and re-structure the prior theory of Zettl, [**27**], and of Everitt and Zettl, [**14**] and [**15**], dealing with multi-interval quasi-differential systems; in particular we relate this analysis of self-adjoint operators to our new version of the GKN-theory involving complex symplectic spaces. Section 7 is devoted to the application of the prior ideas and methods of this paper, to finite multi-interval quasi-differential systems; in particular to the extension of the results in [**9**] and [**10**] for single-interval systems. Finally in Section 8 we develop some general results for the case of infinite dimensional systems, where $\text{card}(\Omega) \geq \aleph_0$, relating to the construction of complete Lagrangian subspaces, and we illustrate these results through several kinds of examples.

2. Some definitions for multi-interval systems

Definition 2.1. *A multi-interval quasi-differential system* (1.1), where Ω is a prescribed (non-empty) *index set* of arbitrary cardinality (within ZFC - set theory), consists of the following data (a) (b) (c):

(a) $\{I_r : r \in \Omega\}$ a set of *real intervals* with $I_r \subseteq \mathbb{R}$, with corresponding endpoints $-\infty \leq a_r < b_r \leq +\infty$, for each $r \in \Omega$. Here each real interval I_r is non-degenerate (contains an interior point), but otherwise I_r may be open, closed, half-open, either compact or non-compact, bounded or unbounded in length. Also the various intervals may be pointwise disjoint, overlapping, or even identical, and the subsets of intervals may be arranged in any possible pattern on $\mathbb{R}$. It is often convenient to consider all the intervals I_r as lying on a single real line $\mathbb{R}$ (see examples in Section 8 below). However, more technically, we consider the set consisting of many piecewise disjoint copies of the real line $\mathbb{R}$, indexed by $r \in \Omega$, and then take I_r as a subinterval of the r-th copy of $\mathbb{R}$. Hence $L^2_{r_1}$ is disjoint from $L^2_{r_2}$ for $r_1 \neq r_2$ in Ω regardless of the positions of I_{r_1} and I_{r_2} on their respective real lines.

(b) $\{M_r : r \in \Omega\}$ a set of *linear ordinary differential expressions* (or formal differential operators); in the *classical* case M_r is of finite order $n_r \geq 1$ on I_r, for each $r \in \Omega$; in the *quasi-differential* case $M_r := M_{A_r}$ is specified by a $n_r \times n_r$ Shin-Zettl matrix A_r on I_r, with $n_r \geq 2$, for each $r \in \Omega$.

In the classical case M_r has complex coefficients and is of the form

$$M_r[y] = p_{n_r} y^{(n_r)} + p_{n_r-1} y^{(n_r-1)} + \cdots + p_1 y' + p_0 y \tag{2.1}$$

where (here $\mathbb{C}$ is the complex number field)

$$p_j : I_r \to \mathbb{C} \text{ with } p_j \in L^1_{\text{loc}}(I_r) \text{ for } j = 0, 1, \cdots, n_r - 1, n_r \tag{2.2}$$

and further

$$p_{n_r} \in AC_{\text{loc}}(I_r) \text{ with } p_{n_r}(x) \neq 0 \text{ for almost all } x \in I_r. \tag{2.3}$$

For the special case $n_r = 1$ see details in [**8**].

In the more general quasi-differential case the expression M_r is defined as in [**15**], [**8**] and [**9**, Section I]; for $n_r \geq 2$ the expression $M_r := M_{A_r}$ is specified by a complex Shin-Zettl matrix A_r of size n_r with

$$M_{A_r}[y] = i^{n_r} y^{[n_r]}_{A_r}, \tag{2.4}$$

where the quasi-derivatives $y^{[j]}_{A_r}$ for $j = 0, 1, 2, \cdots, n_r$ are taken relative to the matrix A_r (as described in the notations of [**15**], [**8**], [**9**, Section I] and [**23**] by $A_r \in Z_{n_r}(I_r)$).

It is known, see [**12**], [**9**, Section I, and Appendices A and B] and [**10**], that every classical ordinary linear differential expression M, as in (b) (2.1) above, can be written as a quasi-differential expression M_A, as in (2.4), with the same order $n \geq 2$. The first order differential expressions are essentially classical in form; see [**8**].

Hence we assume that for each M_r in the multi-interval quasi-differential system (1.1), for all cases when $n_r \geq 2$, is a quasi-differential expression specified by an appropriate Shin-Zettl matrix $A_r \in Z_{n_r}(I_r)$. When $n_r = 1$ we take M_r as a classical expression and the methods given here work also in this case. However to avoid too many details we use only the quasi-differential notation.

Furthermore we assume also that each M_r is a *Lagrange symmetric* (that is, *formally self-adjoint*) expression so that the Lagrange adjoint of M_r is

$$M_r^+ := M_{A_r^+} \quad \text{and} \quad M_r^+ = M_r \text{ for each } r \in \Omega. \tag{2.5}$$

This is equivalent to assuming, in accord with our standing hypothesis, that

$$A_r^+ := -\Lambda_{n_r}^{-1} A_r^* \Lambda_{n_r} \quad \text{and} \quad A_r^+ = A_r \text{ for each } r \in \Omega. \tag{2.6}$$

(Here $A^* = \overline{A^t}$ denotes the conjugate transpose of A, as usual, and Λ_n, for all $n \geq 2$, is the particular $n \times n$ matrix specified in [**28**], [**5**], [**15**], [**12**] and [**9**], *i.e.*

$$\Lambda_n = (l_{st}) = (-1)^s \text{ for } s + t = n + 1, \text{ and } l_{st} = 0 \text{ otherwise.})$$

The corresponding *domain* for $M_r = M_{A_r}$ is denoted by

$$D(A_r) := \{y : I_r \to \mathbb{C} : y^{[j-1]}_{A_r} \in AC_{\text{loc}}(I_r) \quad \text{for} \quad j = 1, 2, \cdots, n_r\}, \tag{2.7}$$

so that $y^{[n_r]}_{A_r}$ and also $M_{A_r}[y]$ both belong to $L^1_{\text{loc}}(I_r)$ for all $y \in D(A_r)$. Then for all pairs $f, g \in D(A_r)$ the *Green's Formula* holds

$$\int_\alpha^\beta \{M_r[f]\overline{g} - f\overline{M_r[g]}\}\, dx = [f, g]_r(\beta) - [f, g]_r(\alpha) \tag{2.8}$$

for every compact interval $[\alpha, \beta] \subset \text{interior}(I_r)$. Here, for all $f, g \in D(A_r)$,

$$[f, g]_r(\cdot) : I_r \to \mathbb{C} \tag{2.9}$$

where, for all $x \in I_r$,

$$[f,g]_r(x) = i^{n_r} \sum_{s=1}^{n_r} (-1)^{s-1} f_{A_r}^{[n_r-s]}(x) \overline{g_{A_r}^{[s-1]}}(x) \tag{2.10}$$

is the quasi-differential representation of the classical sesquilinear form, depending on $f, g \in D(A_r)$ and evaluated at $x \in I_r$; see [**15**], [**8**]. It is then clear that $[f,g]_r(x) = 0$ for x belonging to the complement of supp$(f) \cap$ supp(g) in I_r.

(c) $\{w_r : r \in \Omega\}$ a set of *positive weight functions*, that is,

$$w_r : I_r \to \mathbb{R},\ w_r \in L^1_{\text{loc}}(I_r) \text{ and } w_r(x) > 0 \text{ for almost all } x \in I_r. \tag{2.11}$$

In terms of such a weight function w_r (or measure $w_r\, dx$ on I_r) we introduce the usual *complex Hilbert function space* L^2_r,

$$L^2_r \equiv L^2(I_r; w_r) \tag{2.12}$$

consisting of all functions $f : I_r \to \mathbb{C}$ (or appropriate equivalence classes of functions) for which

$$\int_{I_r} |f|^2 w_r \equiv \int_{I_r} |f(x)|^2 w_r(x)\, dx < +\infty. \tag{2.13}$$

In this space we denote the customary scalar product in L^2_r by

$$\langle f, g \rangle_r = \int_{I_r} f \overline{g} w_r \quad \text{for all} \quad f, g \in L^2_r \tag{2.14}$$

so that the norm in L^2_r is specified by

$$\|f\|_r^2 = \langle f, f \rangle_r \ \text{ for all } f \in L^2_r. \tag{2.15}$$

These definitions and notations hold for all $r \in \Omega$.

For the general theory of Hilbert function spaces see the standard texts [**1**], [**4**, Chapter X - XIII and Appendix], [**24**], [**26**].

Note that for the choice $w_r(x) = 1$ on I_r, $L^2_r = L^2_r(I_r; w_r)$ is the classical complex Hilbert space.

These properties (a) (b) (c) complete the definition, and the standing hypotheses, which apply for the multi-interval system (1.1). However we shall augment Definition 2.1 with the introduction in (d) below of the maximal and minimal operators generated by (1.1) in L^2_r, and the related structure for each $r \in \Omega$. Here summarizing the complete Definition 2.1 for (a) (b) (c) (d), we state that for each fixed $r \in \Omega$ the multi-interval system $\{I_r, M_r, w_r\}$ reduces to a single-interval system as in [**8**], [**9**] and in earlier works, as reviewed briefly in Section 4 below. However the multi-interval system also defines other operators in the collective Hilbert space $\mathbf{H} = \sum_{r \in \Omega} \oplus L^2_r$, as in [**15**] when Ω is either finite or countably denumerable, *i.e.* when $\Omega = \mathbb{N} = \{1, 2, 3, \cdots\}$, and as is considered later in Section 5 for noncountable Ω of higher cardinality $> \aleph_0$.

(d) For details of the definitions and properties given below in this section see [**1**, Appendix II], [**4**, Chapter XII.4, XIII.2] and, in particular, [**24**, Chapter V].

Now we define for all $r \in \Omega$, the *maximal domain* for $w_r^{-1} M_r$ as an operator on L^2_r, namely

$$D_{\max,r} := \{f \in D(A_r) : f \text{ and } w_r^{-1} M_r[f] \in L^2_r\} \tag{2.16}$$

and, accordingly the *maximal operator* $T_{1,r}$ defined, with its domain, by

$$D(T_{1,r}) := D_{\max,r}$$

(2.17) $$T_{1,r}f := w_r^{-1}M_r[f] \text{ for all } f \in D(T_{1,r}).$$

On each domain $D(T_{1,r})$ we introduce the *skew-hermitian sesquilinear form* $[\cdot : \cdot]_r$, namely

(2.18) $$[\cdot : \cdot]_r : D(T_{1,r}) \times D(T_{1,r}) \to \mathbb{C}$$

where, for all $f, g \in D(T_{1,r})$,

(2.19) $$\begin{aligned}[f : g]_r &:= \int_{I_r} \{M_r[f]\overline{g} - f\overline{M_r[g]}\} \\ &= \langle w_r^{-1}M_r[f], g\rangle_r - \langle f, w_r^{-1}M_r[g]\rangle_r \\ &= \langle T_{1,r}f, g\rangle_r - \langle f, T_{1,r}g\rangle_r .\end{aligned}$$

Thus from the Green's formula we obtain, for all $f, g \in D(T_{1,r})$,

$$[f : g]_r = [f, g]_r(b_r^-) - [f, g]_r(a_r^+)$$

(where the last two evaluations at the endpoints $a_r < b_r$ of I_r are each defined by limits that necessarily exist and are finite for all $f, g \in D(T_{1,r})$).

For each index $r \in \Omega$ we have defined the maximal operator $T_{1,r}$ on $D(T_{1,r}) \subset L_r^2$, as generated by $w_r^{-1}M_r$; we now define, for each $r \in \Omega$, the *minimal operator* $T_{0,r}$ on $D(T_{0,r}) \subset L_r^2$. Namely define

(2.20) $$T_{0,r}f := w_r^{-1}M_r[f] \text{ for all } f \in D(T_{0,r})$$

where the *minimal domain* is defined by

(2.21) $$D(T_{0,r}) := \{f \in D(T_{1,r}) : [f : D(T_{1,r})]_r = 0\}.$$

It is known [**15**], [**9**] that $T_{0,r}$ is the minimal symmetric closed operator in L_r^2 which is generated by $w_r^{-1}M_r$ in the space L_r^2. In more detail, if $T_{0\,0,r}$ is generated by $w_r^{-1}M_r$ on the *classical domain*

(2.22)
$$D_0(T_{1,r}) := \{f \in D(T_{1,r}) : \text{ supp}(f) \text{ is a compact set in the interior of } I_r\},$$

which is known to be dense in L_r^2, and thereon

$$T_{0\,0,r}f := w_r^{-1}M_rf \text{ for all } f \in D_0(T_{1,r}),$$

then $T_{0,r}$ is the closure of the operator $T_{0\,0,r}$ in the space L_r^2.

A fundamental theorem of von Neumann [**4**, Chapter XII.4.10] asserts that there is a direct sum decomposition

$$D(T_{1,r}) = D(T_{0,r}) \dotplus N_r^- \dotplus N_r^+,$$

where the deficiency spaces

$$N_r^\pm := \{f \in D(T_{1,r}) : T_{1,r}f = \pm if\}$$

are each Hilbert subspaces of L_r^2. Further the corresponding deficiency indices d_r^-, d_r^+ of $T_{0,r}$ in L_r^2 are calculated from

$$d_r^\pm := \dim(N_r^\pm),$$

respectively. We note that, for $\lambda \in \mathbb{C}$, $T_{1,r}f = \lambda f$ if and only if $M_r[f] = \lambda w_r f$, and hence for each $r \in \Omega$

(2.23) $$0 \le d_r^-, d_r^+ \le n_r.$$

These deficiency indices are important in the theory of self-adjoint operators generated by $w_r^{-1}M_r$ in L_r^2, since such operators exist if and only if

$$d_r^- = d_r^+ \tag{2.24}$$

and their common value is then denoted by $d_r = d_r^{\pm}$.

This completes Definition 2.1 with the parts (a),(b),(c) and (d).

In the single interval case the maximal and minimal operators $T_{1,r}$ and $T_{0,r}$ are known to possess the following fundamental properties; these properties lead directly to the proof of the GKN-Theorem, see [**24**, Chapter V], [**15**] and, in particular, [**8**] and [**9**]. Namely the following results hold for each $r \in \Omega$;

$$T_{0,r} \subseteq T_{1,r} \quad \text{on} \quad D(T_{0,r}) \subseteq D(T_{1,r}) \subset L_r^2 \tag{2.25}$$

so that $T_{0,r}$ is a restriction of $T_{1,r}$, and $T_{1,r}$ is an extension of $T_{0,r}$;

$$D(T_{0,r}),\ \text{and hence}\ D(T_{1,r}),\ \text{are each dense in}\ L_r^2; \tag{2.26}$$

the operators enjoy the adjoint property (with respect to L_r^2)

$$T_{0,r}^* = T_{1,r} \quad \text{and} \quad T_{1,r}^* = T_{0,r}. \tag{2.27}$$

Thus both $T_{0,r}$ and $T_{1,r}$ are closed operators in L_r^2, and furthermore $T_{0,r}$ is a symmetric operator on $D(T_{0,r})$. (Note that the second equality in (2.27) is a consequence of the general theory of symmetric operators, see [**1**] or [**26**], but we include it for essential reference.)

We shall be concerned with self adjoint extensions T_r with domains $D(T_r) \subset L_r^2$, as generated by $w_r^{-1}M_r$ (that is, T_r is an extension of $T_{0,r}$) and this means

$$T_r = T_r^* \quad \text{on} \quad D(T_r) = D(T_r^*). \tag{2.28}$$

From the general theory of adjoints for closed linear operators in Hilbert spaces, it follows that

$$T_{0,r} \subseteq T_r \subseteq T_{1,r} \quad \text{on} \quad D(T_{0,r}) \subseteq D(T_r) \subseteq D(T_{1,r}) \tag{2.29}$$

for each such self-adjoint operator T_r, which thus justifies the terminology for the minimal and maximal operators, $T_{0,r}$ and $T_{1,r}$, respectively.

3. Complex symplectic spaces

We next recall the basic concepts of complex symplectic geometry, see [**9**, Sections I and III] and [**10**], and then show their applications to boundary value theory for single-interval and multi-interval problems, see again (1.1); for example the quotient space $D(T_{1,r})/D(T_{0,r})$ is known to be a complex symplectic space for each $r \in \Omega$; compare with (1.2).

Remark on notation. We shall normally use the font sans serif for all vectors and linear manifolds in complex symplectic spaces, whether these spaces are defined for single-interval or multi-interval systems; see Notations 4.1 below.

Definition 3.1. A *complex symplectic space* S is a complex linear space, with a prescribed *symplectic form* $[\cdot : \cdot]$; namely, a sesquilinear form

$$[\cdot : \cdot] : \mathsf{S} \times \mathsf{S} \to \mathbb{C} \tag{3.1}$$

with the properties

$$[c_1\mathsf{u} + c_2\mathsf{v} : \mathsf{w}] = c_1[\mathsf{u} : \mathsf{w}] + c_2[\mathsf{v} : \mathsf{w}] \tag{3.2}$$

(3.3) $$[\mathsf{u} : c_1\mathsf{v} + c_2\mathsf{w}] = \overline{c_1}[\mathsf{u} : \mathsf{v}] + \overline{c_2}[\mathsf{u} : \mathsf{w}]$$

(3.4) $$[\mathsf{u} : \mathsf{v}] = -\overline{[\mathsf{v} : \mathsf{u}]}$$

(3.5) $$[\mathsf{u} : \mathsf{S}] = 0 \quad \text{implies} \quad \mathsf{u} = 0$$

for all vectors $\mathsf{u}, \mathsf{v}, \mathsf{w} \in \mathsf{S}$ and all scalars $c_1, c_2 \in \mathbb{C}$.

Remark 3.1.

1. It is clear that (3.2) and (3.4) imply (3.3).
2. The condition (3.5) is required to ensure that $[\cdot : \cdot]$ is not degenerate on the vector space S. However, sometimes it is convenient to refer to a degenerate "symplectic product" satisfying only (3.1) to (3.4) on a complex vector space, but these four conditions alone do not define a complex symplectic space unless the additional axiom (3.5) also holds.

Definition 3.2. Let S be a complex linear space with a prescribed symplectic product $[\cdot : \cdot]$ (possibly degenerate). Assume that S is represented as a finite direct sum of the linear submanifolds $\{\mathsf{S}_r : r = 1, 2, \ldots, N\}$ so that

$$\mathsf{S} = \mathsf{S}_1 \dotplus \mathsf{S}_2 \dotplus \cdots \dotplus \mathsf{S}_N.$$

That is, each vector $\mathsf{v} \in \mathsf{S}$ has a unique representation as a sum in S

$$\mathsf{v} = \sum_{r=1}^{N} \mathsf{v}_r, \text{ with } \mathsf{v}_r \in \mathsf{S}_r \text{ for } r = 1, 2, \ldots, N.$$

Then the notation

$$\mathsf{S} = \sum_{r=1}^{N} \oplus \mathsf{S}_r, \text{ or } \mathsf{S} = \mathsf{S}_1 \oplus \mathsf{S}_2 \oplus \cdots \oplus \mathsf{S}_N,$$

means that the summands are pairwise symplectically orthogonal; that is

$$[\mathsf{S}_r : \mathsf{S}_t] = 0 \text{ for } 1 \le r \ne t \le N.$$

In particular, if S is a complex symplectic space, so $[\cdot : \cdot]$ is nondegenerate on S, then each S_r, with $[\cdot : \cdot]$, is also a complex symplectic space.

Because of the analogy of the complex symplectic space S to a hermitian scalar-product space, we often employ the terminology *symplectic product* for $[\cdot : \cdot]$, and *symplectic orthogonality* for the pair of vectors u, v when $[\mathsf{u} : \mathsf{v}] = 0$. Hence condition (3.5) above means that only the zero vector of S is symplectically orthogonal to every vector of S.

In this sense we often refer to the geometry and linear algebra of complex symplectic spaces as (linear) symplectic geometry and algebra, especially when S has finite (complex) dimension. See [**9**, Appendix B] and [**22**] for discussions of the corresponding structures of real symplectic spaces and their applications to Lagrange-Hamilton classical mechanics.

As is customary, we declare that the complex symplectic spaces S_1, with form $[\cdot : \cdot]_1$, and S_2, with form $[\cdot : \cdot]_2$, are isomorphic in case there exists a bijective linear map F on S_1 onto S_2,

(3.6) $$F : \mathsf{S}_1 \to \mathsf{S}_2 \quad \text{with} \quad [\mathsf{u} : \mathsf{v}]_1 = [F\mathsf{u} : F\mathsf{v}]_2$$

for all vectors $\mathsf{u}, \mathsf{v} \in \mathsf{S}_1$.

Linear manifolds of a complex symplectic space S need not be complex symplectic spaces, since the induced complex symplectic form can be degenerate on such a manifold.

Definition 3.3. A linear submanifold L in the complex symplectic space S is called *Lagrangian* in case:

$$[\mathsf{L} : \mathsf{L}] = 0, \tag{3.7}$$

that is $[\mathsf{u} : \mathsf{v}] = 0$ for all vectors $\mathsf{u}, \mathsf{v} \in \mathsf{L}$. Further, a Lagrangian manifold $\mathsf{L} \subset \mathsf{S}$ is *complete* in case

$$\mathsf{u} \in \mathsf{S} \text{ and } [\mathsf{u} : \mathsf{L}] = 0 \text{ imply } \mathsf{u} \in \mathsf{L}. \tag{3.8}$$

We refer to a *linear manifold* - more general than a *linear subspace* which is usually required to be a closed subset in some appropriate topology - in the case where S has infinite dimension (that is, S is not finite dimensional).

Example 3.1. Construct an infinite dimensional complex linear space $\mathsf{S} = H_- \oplus H_+$ as the direct sum of two Hilbert spaces H_- and H_+ (say at least one of infinite dimension) so that each vector $\mathsf{u} \in \mathsf{S}$ has a unique representation $\mathsf{u} = \{u_-, u_+\}$ with $u_\pm \in H_\pm$, respectively. In this situation we can define a symplectic form on S by

$$[\mathsf{u} : \mathsf{v}] := -i \langle u_-, v_- \rangle_- + i \langle u_+, v_+ \rangle_+ \tag{3.9}$$

in terms of the hermitian or scalar products $\langle \cdot, \cdot \rangle_\pm$ in the complex Hilbert spaces $H_\pm$, respectively, and S is then a complex symplectic space of infinite dimension. Then $H_\pm$ can be regarded as linear manifolds $\mathsf{H}_\pm$ of S; in fact they can be regarded as symplectic subspaces, closed in the Hilbert space product topology on S and also in other weaker topologies to be considered in Section 6.

Consider in more detail the above defined space S with symplectic form given by (3.9). In particular, for vectors u, v as above, and now holding to our notation for symplectic spaces so that $\mathsf{u}_\pm$ are to be identified with $u_\pm$,

$$[\mathsf{u}_- : \mathsf{v}_-] := -i \langle u_-, v_- \rangle_- \quad \text{for all } u_-, v_- \in H_-$$
$$[\mathsf{u}_+ : \mathsf{v}_+] := +i \langle u_+, v_+ \rangle_+ \quad \text{for all } u_+, v_+ \in H_+.$$

Also note that

$$[\mathsf{H}_- : \mathsf{H}_+] = 0 \text{ and } \mathsf{S} = \mathsf{H}_- \oplus \mathsf{H}_+,$$

so that H_- and H_+ are *symplectic orthocomplements* in the symplectic space S. In addition we can also treat S as a Hilbert space with norm $\|\cdot\|_{\mathsf{S}}$ given by

$$\|\mathsf{u}\|_{\mathsf{S}}^2 = \|\mathsf{u}_-\|_-^2 + \|\mathsf{u}_+\|_+^2 = \|u_-\|_-^2 + \|u_+\|_+^2 .$$

Then H_- and H_+ are metric orthocomplements, and hence closed subspaces, of the Hilbert space S.

PROPOSITION 3.1. *Let $\mathsf{S} = H_- \oplus H_+$ be a complex symplectic space, where $H_\pm$ are each Hilbert spaces, as above. Also S is a Hilbert space, as the direct sum of H_- and H_+, and we use the corresponding metric topology in S; further we can regard $H_\pm$ as symplectic subspaces of S and then use the notation $\mathsf{H}_\pm$.*

Then the quantities

$$\mathsf{u}_+ \in \mathsf{H}_+,\ \mathsf{u}_- \in \mathsf{H}_-,\ \langle \mathsf{u}_\pm, \mathsf{u}_\pm \rangle_\pm,\ \langle \mathsf{u}, \mathsf{v} \rangle \text{ and } [\mathsf{u} : \mathsf{v}]$$

as well as $c_1\mathsf{u} + c_2\mathsf{v}$ in S, all depend continuously on $(\mathsf{u}, \mathsf{v}, c_1, c_2) \in \mathsf{S} \times \mathsf{S} \times \mathbb{C}^2$.

Moreover the closure of a Lagrangian submanifold of S *is itself a Lagrangian subspace. Furthermore, each complete Lagrangian submanifold of* S *is a closed linear subspace of* S.

PROOF. The assertions of continuity are obvious. We shall comment only on the closure $\overline{\mathsf{L}}$ of a complete Lagrangian submanifold $\mathsf{L} \subset \mathsf{S}$. Clearly $\overline{\mathsf{L}}$ is a subspace of the Hilbert space $\mathsf{S} = H_- \oplus H_+$. Now take any point $\mathsf{u} \in \overline{\mathsf{L}}$. Then there exists a sequence of vectors $\{\mathsf{u}^{(k)} \in \mathsf{L} : k \in \mathbb{N}\}$ with $\lim_{k\to\infty} \mathsf{u}^{(k)} = \mathsf{u}$. For each $\mathsf{v} \in \mathsf{L}$, we have $[\mathsf{u}^{(k)} : \mathsf{v}] = 0$. So $[\mathsf{u} : \mathsf{v}] = 0$ and hence $[\mathsf{u} : \mathsf{L}] = 0$; therefore $\mathsf{u} \in \mathsf{L}$ and $\mathsf{L} = \overline{\mathsf{L}}$ is closed. □

Now consider any complex symplectic space S of dimension $D \geq 1$ (the case $D = 0$ is the trivial space consisting of a single point only, and this case will often be omitted from subsequent discussions). It is evident that in any such complex symplectic space S, with symplectic form $[\cdot : \cdot]$, each vector $\mathsf{v} \in \mathsf{S}$ satisfies $\mathrm{Re}[\mathsf{v} : \mathsf{v}] = 0$. Hence v, in fact each 1-dimensional subspace $\mu\mathsf{v}$, for $\mu \in \mathbb{C}$, whereon

$$[\mu\mathsf{v} : \mu\mathsf{v}] = \mu\bar{\mu}[\mathsf{v} : \mathsf{v}] = |\mu|^2 [\mathsf{v} : \mathsf{v}], \tag{3.10}$$

is exactly one of the following three types

$$\begin{array}{lll} (i) & \text{positive} & \mathrm{Im}[\mathsf{v} : \mathsf{v}] > 0 \\ (ii) & \text{negative} & \mathrm{Im}[\mathsf{v} : \mathsf{v}] < 0 \\ (iii) & \text{neutral} & \mathrm{Im}[\mathsf{v} : \mathsf{v}] = 0, \text{ so } [\mathsf{v} : \mathsf{v}] = 0. \end{array} \tag{3.11}$$

Hence a Lagrangian submanifold $\mathsf{L} \subset \mathsf{S}$ consists of neutral vectors, such that $[\mathsf{u} : \mathsf{v}] = 0$ for all $\mathsf{u}, \mathsf{v} \in \mathsf{L}$.

We shall use these ideas to obtain symplectic invariants for complex symplectic spaces S of finite (complex) dimension $D \geq 1$, and some of these invariants will continue to be meaningful (in appropriate interpretations) for infinite dimensional spaces, see Section 6.

Definition 3.4. In a complex symplectic space S with symplectic form $[\cdot : \cdot]$ and *finite dimension* $D \geq 1$, define the following symplectic invariants of S (see [**9**, Section III.1]):

$$\begin{cases} p := \max\{(\text{complex}) \text{ dimension of linear subspaces whereon} \\ \qquad \mathrm{Im}\,[\mathsf{v} : \mathsf{v}] > 0 \text{ for all } \mathsf{v} \neq 0\} \\ q := \max\{(\text{complex}) \text{ dimension of linear subspaces whereon} \\ \qquad \mathrm{Im}\,[\mathsf{v} : \mathsf{v}] < 0 \text{ for all } \mathsf{v} \neq 0\} \end{cases} \tag{3.12}$$

Here (p, q) is the *signature* of S consisting of the ordered pair of integers, the *positivity index* $p \geq 0$ and the *negativity index* $q \geq 0$. (Note: These definitions are phrased slightly differently from those in [**9**, Section III, (1.17)] but they are equivalent for $\dim(\mathsf{S}) < \infty$). Further define the *Lagrangian index*

$$\Delta := \max\{(\text{complex}) \text{ dimension of Lagrangian subspaces of } \mathsf{S}\}; \tag{3.13}$$

and also the *excess* of S

$$Ex := p - q \tag{3.14}$$

as the excess of positivity over negativity of indices of S.

To complete this Definition 3.4 we note that these symplectic invariants of S

$$p, q, (p, q), Ex, \Delta \tag{3.15}$$

are each defined intrinsically in terms of the symplectic structure of S.

It is known, see [**9**, Section III] and [**10**], that each of the pairs (p, q) or (D, Ex) or (Δ, Ex) yields a complete set of invariants which characterize S up to complex symplectic isomorphism. In particular, S is symplectically isomorphic with $\mathbb{C}^D$ when the symplectic form is specified by the skew-hermitian $D \times D$ matrix $\operatorname{diag}\{iI_p, -iI_q\}$, or alternatively $\operatorname{diag}\{iI_q, -iI_q, iI_{|Ex|}\}$ (where I_q is the notation for the identity matrix of size $q \geq 1$, and the alternative format is indicated for the case $p > q \geq 1$—with obvious modifications in the case $p \leq q$ or $q = 0$). From such a skew-hermitian $D \times D$ matrix, in diagonal format, it is immediately clear that

$$D = p + q, \quad Ex = p - q, \quad \Delta = \min\{p, q\} = \tfrac{1}{2}(D - |Ex|). \tag{3.16}$$

THEOREM 3.1. *Let S be a finite dimensional complex symplectic space, with symplectic product $[\cdot : \cdot]$, and symplectic invariant indices*

$$p \ (positivity) \quad q \ (negativity) \quad \Delta \ (Lagrangian).$$

Assume that there exists a symplectically orthogonal direct sum decomposition into finitely many summands,

$$\mathsf{S} = \sum_{r=1}^{N} \oplus \mathsf{S}_r.$$

Then each S_r is a complex symplectic subspace of S, with the corresponding invariants p_r, q_r, Δ_r for $1 \leq r \leq N$.

Furthermore

$$p = \sum_{r=1}^{N} p_r \qquad q = \sum_{r=1}^{N} q_r \qquad \Delta \geq \sum_{r=1}^{N} \Delta_r \tag{3.17}$$

and thus

$$\dim(\mathsf{S}) = \sum_{r=1}^{N} \dim(\mathsf{S}_r) \qquad Ex(\mathsf{S}) = \sum_{r=1}^{N} Ex(\mathsf{S}_r). \tag{3.18}$$

If $Ex(\mathsf{S}_r) = 0$ for $r = 1, 2, \ldots, N$, then both $Ex(\mathsf{S}) = 0$ and $\Delta = \sum_{r=1}^{N} \Delta_r$; and conversely.

PROOF. If a vector $\mathsf{f} \in \mathsf{S}_1$ is symplectically orthogonal to all S_1, that is $[\mathsf{f} : \mathsf{S}_1] = 0$, then $[\mathsf{f} : \mathsf{S}] = 0$ and so $\mathsf{f} = 0$. Hence S_1, and similarly each S_r for $r = 1, 2, \ldots, N$, inherits a non-degenerate symplectic product from S, and therefore each S_r is a complex symplectic subspace of S.

From the description of S_r as a complex vector space of dimension $\dim(\mathsf{S}_r)$, and noting the corresponding skew-hermitian matrix in diagonal format, as above, we immediately calculate the desired equalities for p and q; the other results then follow - compare [**9**, Section 3.1, Theorem 1 and Corollaries]. For instance we prove the last assertion of the theorem.

If $Ex(\mathsf{S}_r) = 0$ for $r = 1, 2, \ldots, N$, then $p_r = q_r = \Delta_r$; hence $Ex(\mathsf{S}) = p - q = 0$ and $\Delta = p = q = \sum_{r=1}^{N} \Delta_r$. Conversely assume that $Ex(\mathsf{S}) = 0$ and $\Delta = \sum_{r=1}^{N} \Delta_r$. Suppose some $Ex(\mathsf{S}_r) \neq 0$, say $p_1 > q_1$ without loss of generality. In such a case

$$\Delta = \sum_{r=1}^{N} p_r > q_1 + \sum_{r=2}^{N} \Delta_r = \sum_{r=1}^{N} \Delta_r,$$

which is a contradiction. □

4. Single interval quasi-differential systems

Now we review the application of complex symplectic algebra to the analysis of a single-interval system $\{I, M_A, w\}$, involving a quasi-differential expression M_A, for complex Shin-Zettl matrix $A = A^+ \in Z_n(I)$ of size $n \geq 2$ on the real interval I, and with reference to the positive weight function w specifying the complex Hilbert space $L^2(I; w)$. In the single interval case, where $\Omega = \{1\}$, we simplify the notation of Section 2 (see Notations 4.1 below) by suppressing the unnecessary index $r = 1$, in accord with [**9**, Section III.2] and earlier works.

LEMMA 4.1. *For the single system* $\{I, M_A, w\}$ *let* T_0 *on* $D(T_0)$ *and* T_1 *on* $D(T_1)$ *be the minimal and maximal operators, respectively, as generated by* $w^{-1}M_A$ *in* $L^2(I; w)$. *Then the quotient (or identification) space*

$$\mathsf{S} := D(T_1)/D(T_0) \tag{4.1}$$

is a complex symplectic space, with $\dim(\mathsf{S}) \leq 2n$, *where the symplectic form is specified by*

$$[\mathsf{f} : \mathsf{g}] := [f : g]_A = \langle T_1 f, g\rangle - \langle f, T_1 g\rangle \quad \textit{for all} \quad f, g \in D(T_1). \tag{4.2}$$

PROOF. See [**9**, Section III.2], where we note that the symplectic product $[\mathsf{f} : \mathsf{g}]$ does not depend on the representatives f and g of the respective cosets. □

Notations 4.1. For a single interval system, denoted by $\{I, M_A, w\}$ for $\Omega = \{1\}$, we write A (for A_1), $L^2(I; w)$ (for $L^2(I_1; w_1)$) with scalar product $\langle\cdot,\cdot\rangle$ (for $\langle\cdot,\cdot\rangle_1$), the corresponding minimal and maximal operators T_0 and T_1, respectively, and the symplectic product $[\cdot : \cdot]_A$ (for $[\cdot : \cdot]_1$) - as compared with Section 2 above.

We call S the *boundary* or *end-point space* for $\{I, M_A, w\}$, and we introduce the notation

$$\mathsf{f} = \{f + D(T_0)\} \quad \text{or} \quad \mathsf{f} = \Psi f, \tag{4.3}$$

similarly $\mathsf{g} = \{g + D(T_0)\}$, *etc.*, where f is the coset of $f \in D(T_1)$, and so it is the element of S which is the image of f under the natural projection Ψ of $D(T_1)$ onto S:

$$\Psi : D(T_1) \to \mathsf{S}, \quad f \to \Psi f = \mathsf{f} = \{f + D(T_0)\}. \tag{4.4}$$

We especially remark on the notation where complex functions f, g, h, u, v, w, *etc.*, in Hilbert spaces, say $L^2(I; w)$ or H, are in *mathematical italic* font; whereas $\mathsf{f} = \{f + D(T_0)$ and $\mathsf{g} = \{g + D(T_0)\}$, and also $\mathsf{h}, \mathsf{u}, \mathsf{v}, \mathsf{w}$, *etc.*, are in sans serif font, to indicate cosets, as are spaces of cosets L and H of symplectic spaces $\mathsf{S} = D(T_1)/D(T_0)$ (see Proposition 3.1 above and also Remark 5.1 below).

These notational differences will be used systematically as a convenience to the reader, but, in every case, the symbol is appropriately and clearly defined without reference to the printing style or font.

We now assert the GKN-Theorem (Glazman-Krein–Naimark) for the single-interval system $\{I, M_A, w\}$; see the early form of this result in [**24**, Chapter V], and for the complete result in [**8**] and [**9**, Section III.2].

THEOREM 4.1 (Glazman-Krein-Naimark). *Consider the single-interval system* $\{I, M_A, w\}$ *in the space* $L^2(I;w)$, *with* $A = A^+ \in Z_n(I)$; *let* T_0 *on* $D(T_0)$ *and* T_1 *on* $D(T_1)$ *be, respectively, the minimal and maximal operators; assume that* T_0 *has equal deficiency indices* $d := d^- = d^+$ *where* $0 \leq d \leq n$.

As in Lemma 4.1 *above let*

$$\mathsf{S} := D(T_1)/D(T_0) \tag{4.5}$$

be the endpoint complex symplectic space, with $\dim(\mathsf{S}) \leq 2n$.

Then there is a natural bijection of the set $\{T\}$ *of all self-adjoint extensions of* T_0 *in* $L^2(I;w)$, *and the set* $\{\mathsf{L}\}$ *of all complete Lagrangians of* S. *Namely for each such self-adjoint operator* T *on* $D(T) \subset L^2(I;w)$ *the corresponding complete Lagrangian subspace* $\mathsf{L} \subset \mathsf{S}$ *is*

$$\mathsf{L} := \Psi D(T) = D(T)/D(T_0) \tag{4.6}$$

where the natural projection Ψ *is specified in* (4.4).

Furthermore

$$D(T) = \{f \in D(T_1) : [\mathsf{f} : \mathsf{L}] = 0\} \tag{4.7}$$

or equally well,

$$D(T) = c_1 f_1 + \ldots + c_d f_d + D(T_0) \tag{4.8}$$

where the functions $f_1, \cdots, f_d$ *of* $D(T_1)$ *have images* $\{\mathsf{f}_1, \ldots, \mathsf{f}_d\}$ *that are a basis for* L, *and* $c_1, \ldots, c_d$ *are arbitrary complex numbers. Moreover, for each* $g \in D(T)$ *there exist unique complex numbers* $c_1, c_2, \ldots, c_d$ *and a function* $g_0 \in D(T_0)$ *so that*

$$g = c_1 f_1 + c_2 f_2 + \cdots + c_d f_d + g_0.$$

PROOF. See [**9**, Section III] or [**10**]. □

Remark 4.1. By this version of the GKN-Theorem there exists a self-adjoint extension of T_0 in $L^2(I;w)$, as generated by $w^{-1}M_A$, if and only if there exists a complete Lagrangian subspace $\mathsf{L} \subset \mathsf{S}$. This is the case if and only if the excess Ex of S is zero; in this case a Lagrangian subspace is complete if and only if it has dimension one half the dimension of S.

The invariants of the complex symplectic space S can be expressed in terms of the deficiency indices $d^{\pm}$ of the closed symmetric operator T_0, defined by, recalling that $T_1 = T_0^*$,

$$d^{\pm} := \dim\{f \in D(T_1) : T_1 f = \pm i f\} \tag{4.9}$$

and then

$$p = d^+,\ q = d^- \text{ so } \dim(\mathsf{S}) = d^+ + d^-,\ Ex = d^+ - d^-. \tag{4.10}$$

Thus a necessary and sufficient condition that $\{T\}$, and hence $\{\mathsf{L}\}$, be non-empty is

$$Ex = d^+ - d^- = 0, \text{ and then we can write } d = d^{\pm} \text{ and } \dim(\mathsf{S}) = 2d. \tag{4.11}$$

In this case a Lagrangian subspace $\mathsf{L} \subset \mathsf{S}$ is complete if and only if $\dim(\mathsf{L}) = d = \frac{1}{2}\dim(\mathsf{S})$. In the special case where $T_0 = T_1$, so that $\mathsf{S} = 0$, the unique self-adjoint operator T on $D(T)$ (which corresponds to $\mathsf{L} = 0$) is just $T = T_0 = T_0^* = T_1$; that is there are no proper self-adjoint extensions of T_0, and similarly no proper self-adjoint restrictions of T_1.

Remark 4.2. For each self-adjoint extension T on $D(T)$ of T_0 on $D(T_0) \subset L^2(I;w)$, the corresponding complete Lagrangian d-space $\mathsf{L} \subset \mathsf{S}$ can be interpreted as the set of linear functionals on $D(T_1) \to \mathbb{C}$ which vanish on $D(T_0)$, and which thereby are the "boundary conditions" whose null space precisely specifies $D(T)$ in more familiar language, see [**9**, Section III].

Remark 4.3. The particular case where the order $n = 1$ is treated in the GKN-theory in the earlier work of the authors [**8**], but is omitted from our subsequent discussions here.

5. Multi-interval quasi-differential systems

Let $\{I_r, M_r, w_r : r \in \Omega\}$ be a multi-interval quasi-differential system as in (1.1) and (a), (b), (c) and (d) of Definition 2.1 of the Section 2. In particular, we recall the assumptions (2.5) to (2.10) that M_r is Lagrange symmetric on the interval I_r, for each $r \in \Omega$. We define a complex Hilbert space $\mathbf{H}$, together with maximal and minimal operators $\mathbf{T}_1$ on $\mathbf{D}(\mathbf{T}_1) \subset \mathbf{H}$ and $\mathbf{T}_0$ on $\mathbf{D}(\mathbf{T}_0) \subset \mathbf{H}$, respectively, in order to provide an operator-theoretic framework for the collective system. Of course, all these concepts and results developed for multi-interval systems also apply to the single systems of Section 4 (take $\Omega = \{1\}$) but new kinds of problems and methods arise here in Section 5.

Remark 5.1 (on notation)**.** In Sections 2 and 4 above, we considered single-interval systems $\{I, M_A, w\}$, as in Theorem 4.1 (GKN), and studied functions $f \in D(T_1) \subset L^2(I;w)$ and the corresponding cosets $\mathsf{f} = \{f + D(T_0)\} \in \mathsf{S}$. We noted the convention that sans serif type denotes these cosets or spaces of such cosets.

Similar notations are used for multi-interval systems $\{I_r, M_r, w_r : r \in \Omega\}$, except we use **bold face** type for system operators and spaces, say

$$\mathbf{f} \in \mathbf{H} \text{ with } \mathsf{f} = \{\mathbf{f} + \mathbf{D}(\mathbf{T}_0)\} \in \mathsf{S};$$

similarly for $\mathsf{g} = \{\mathbf{g} + \mathbf{D}(\mathbf{T}_0)\}$, and also for $\mathsf{h}, \mathsf{u}, \mathsf{v}, \mathsf{w}$, *etc.*, see Notations 4.1 above.

Definition 5.1. Let $\{I_r, M_r, w_r : r \in \Omega\}$ be a multi-interval quasi-differential system. Define the (system) complex Hilbert space $\mathbf{H}$ for this multi-interval system to be

$$\mathbf{H} := \sum_{r \in \Omega} \oplus L_r^2, \tag{5.1}$$

that is, the direct sum (possibly non-countable) of the spaces $L_r^2 \equiv L^2(I_r; w_r)$ with $r \in \Omega$. Namely an element $\mathbf{f}$ of $\mathbf{H}$ is a (generalized) sequence

$$\mathbf{f} = \{f_r : r \in \Omega\} \tag{5.2}$$

with each component $f_r \in L_r^2$, and the usual convention that the function f_r also stands for the corresponding equivalence class. In more detail $\mathbf{f}$ is a function from Ω into the union $\bigcup_{r \in \Omega} L_r^2$, with $r \in \Omega \to f_r \in L_r^2$ and with the norm $\|\cdot\|_{\mathbf{H}} \geq 0$ defined by

$$\|\mathbf{f}\|_{\mathbf{H}}^2 := \sum_{r \in \Omega} \|f_r\|_r^2 < \infty; \tag{5.3}$$

that is, $\mathbf{f} \in \mathbf{H}$ just in case

$$\|\mathbf{f}\|^2_{\mathbf{H}} := \sup \left\{ \sum_{r \in \omega} \|f_r\|^2_r \right\} < \infty, \tag{5.4}$$

where the supremum is taken over all finite subsets ω of elements of Ω. It is clear that for each $\mathbf{f} \in \mathbf{H}$, all but a countable number of components satisfy $f_r = 0$ in L^2_r. In fact for each countable set $\{\mathbf{f}_k \in \mathbf{H} : k \in \mathbb{N}\}$ there exists a countable subset $\Omega_\infty \subseteq \Omega$ such that every component $f_{k,r} \in L^2_r$ of $\mathbf{f}_k$ vanishes for each $k \in \mathbb{N}$, and for each r in the complement of Ω_∞.

Then, when $\mathbf{H}$ is endowed with componentwise linear algebra and a familiar scalar product, we verify that $\mathbf{H}$ is a complex Hilbert space, with a dimension (the cardinality of a complete orthonormal basis) which can be non-countable. Here the scalar product of $\mathbf{f}$ and $\mathbf{g}$ in $\mathbf{H}$ is given by the infinite series

$$\langle \mathbf{f}, \mathbf{g} \rangle_{\mathbf{H}} = \sum_{r \in \Omega} \langle f_r, g_r \rangle_r \,, \tag{5.5}$$

which is absolutely convergent since

$$\sum_{r \in \Omega} |\langle f_r, g_r \rangle_r| \le \sum_{r \in \Omega} (\|f_r\|_r \cdot \|g_r\|_r) \le \frac{1}{2} \sum_{r \in \Omega} \left(\|f_r\|^2_r + \|g_r\|^2_r \right) < \infty, \tag{5.6}$$

and the corresponding norm is given by

$$\|\mathbf{f}\|^2_{\mathbf{H}} = \langle \mathbf{f}, \mathbf{f} \rangle_{\mathbf{H}} = \sum_{r \in \Omega} \|f_r\|^2_r \,. \tag{5.7}$$

Hence $\|\mathbf{f}\|^2_{\mathbf{H}} = 0$ if and only if every $f_r = 0$ in L^2_r for all $r \in \Omega$, and two such generalized sequences $\mathbf{f}$ and $\mathbf{g}$ define the same element in $\mathbf{H}$ if and only if $\|\mathbf{f} - \mathbf{g}\|_{\mathbf{H}} = 0$.

Next we define the maximal and minimal operators $\mathbf{T}_1$ on $\mathbf{D}(\mathbf{T}_1) \subset \mathbf{H}$ and $\mathbf{T}_0$ on $\mathbf{D}(\mathbf{T}_0) \subset \mathbf{H}$, respectively, for the system $\{I_r, M_r, w_r : r \in \Omega\}$.

Definition 5.2. Let $\{I_r, M_r, w_r : r \in \Omega\}$ be a multi-interval system, as defined in Definition 2.1 above, and let the corresponding complex Hilbert space $\mathbf{H}$ be defined as in 5.1 above. Define the *maximal operator* $\mathbf{T}_1$ for this system, on the maximal domain $\mathbf{D}(\mathbf{T}_1) \subset \mathbf{H}$, by

$$\begin{aligned} \mathbf{D}(\mathbf{T}_1) \;:=\; \{\mathbf{f} = \{f_r : r \in \Omega\} \in \mathbf{H} : \; & (i)\ f_r \in D(T_{1,r}) \text{ for all } r \in \Omega \\ & \text{and } (ii)\ \{T_{1,r} f_r : r \in \Omega\} \in \mathbf{H}\} \end{aligned} \tag{5.8}$$

and then

$$\mathbf{T}_1 \mathbf{f} := \{T_{1,r} f_r : r \in \Omega\} \text{ for all } \mathbf{f} \in \mathbf{D}(\mathbf{T}_1). \tag{5.9}$$

Similarly define

$$\mathbf{D}(\mathbf{T}_0) := \{\mathbf{f} = \{f_r : r \in \Omega\} \in \mathbf{D}(\mathbf{T}_1) : f_r \in D(T_{0,r}) \text{ for all } r \in \Omega\} \tag{5.10}$$

and then

$$\mathbf{T}_0 \mathbf{f} := \mathbf{T}_1 \mathbf{f} \text{ for all } \mathbf{f} \in \mathbf{D}(\mathbf{T}_0). \tag{5.11}$$

It is clear that we can interpret L^2_t as a Hilbert subspace of $\mathbf{H}$, for each fixed $t \in \Omega$; namely L^2_t consists of all vectors $\mathbf{f} = \{f_r : r \in \Omega\} \in \mathbf{H}$ with $f_r = 0$ for all r

different from t. Also, in the same way, $D(T_{1,t}) \subset \mathbf{D}(\mathbf{T}_1)$ and $D(T_{0,t}) \subset \mathbf{D}(\mathbf{T}_0)$ for each fixed $t \in \Omega$.

LEMMA 5.1. *Both* $\mathbf{T}_0$ *on* $\mathbf{D}(\mathbf{T}_0)$ *and* $\mathbf{T}_1$ *on* $\mathbf{D}(\mathbf{T}_1)$ *are closed operators with dense domains* $\mathbf{D}(\mathbf{T}_0) \subseteq \mathbf{D}(\mathbf{T}_1)$ *in* $\mathbf{H}$. *In fact,* $\mathbf{T}_0$ *on* $\mathbf{D}(\mathbf{T}_0)$ *is an extension of the operator* $\mathbf{T}_{00}$, *which is defined as the restriction of* $\mathbf{T}_1$ *to the domain* $\mathbf{D}(\mathbf{T}_{00})$, *given by (compare* (2.22))

$$\mathbf{D}(\mathbf{T}_{00}) := \{\mathbf{f} \in \mathbf{D}(\mathbf{T}_1) : \text{only a finite number of components of } \mathbf{f} = \{f_r : r \in \Omega\} \text{ are non-zero, and each of these satisfies } f_r \in D_0(T_{1,r})\}. \tag{5.12}$$

That is, each of the finitely many non-zero component f_r *of* $\mathbf{f} \in \mathbf{D}(\mathbf{T}_{00})$ *has a compact support that lies interior to* I_r. *Furthermore* $\mathbf{D}(\mathbf{T}_{00})$ *(sometimes denoted by* $\mathbf{D}_0(\mathbf{T}_1)$) *is dense in* $\mathbf{H}$.

PROOF. The proof that $\mathbf{T}_0$ and $\mathbf{T}_1$ are closed operators in $\mathbf{H}$ is given in [**15**] for the case when Ω is denumerable. However, this is no significant restriction because the relevant proofs involve only a countable set of elements $\{\mathbf{f}_k = \{f_{k,r} : r \in \Omega\} \in \mathbf{H} : k \in \mathbb{N}\}$, and then there exists a denumerable set of indices, say $\Omega_\infty \subseteq \Omega$, such that for every $k \in \mathbb{N}$ we have $f_{k,r} = 0$ in L^2_r when r is in the complement of Ω_∞.

However, in order to emphasize certain facts needed later, and to illustrate the method for the reduction to a countable set of indices $r \in \Omega$, we here present the proof that $\mathbf{T}_1$ is closed. Let $\{\mathbf{f}_k \in \mathbf{D}(\mathbf{T}_1) : k \in \mathbb{N}\}$ be a sequence such that, for given $\mathbf{f} = \{f_r : r \in \Omega\}$, $\mathbf{g} = \{g_r : r \in \Omega\} \in \mathbf{H}$,

$$\lim_{k\to\infty} \mathbf{f}_k = \mathbf{f} \text{ and } \lim_{k\to\infty} \mathbf{T}_1\mathbf{f}_k = \mathbf{g} \text{ in the metric of } \mathbf{H};$$

then we must show that $\mathbf{f} \in \mathbf{D}(\mathbf{T}_1)$ with $\mathbf{T}_1\mathbf{f} = \mathbf{g}$. Now take a countable subset $\Omega_\infty = \mathbb{N} \subseteq \Omega$ such that all the components $\{f_{k,r}, f_r, T_1 f_{k,r}, g_r\}$ are all zero for $k \in \mathbb{N}$ and $r \in \Omega \setminus \Omega_\infty$.

Then for each $r \in \Omega_\infty$,

$$\|f_{k,r} - f_r\|_r \leq \|\mathbf{f}_k - \mathbf{f}\|_{\mathbf{H}} \text{ for all } k \in \mathbb{N}$$

and so

$$\lim_{k\to\infty} \|f_{k,r} - f_r\|_r = 0 \text{ for each fixed } r \in \Omega_\infty.$$

Similarly $\lim_{k\to\infty} \|T_{1,r} f_{k,r} - g_r\|_r = 0$.

Since $T_{1,r}$ on $D(T_{1,r}) \subset L^2_r$ is a closed operator, for each fixed $r \in \Omega_\infty$, we obtain

$$f_r \in D(T_{1,r}) \text{ and } T_{1,r} f_r = g_r \text{ for all } r \in \Omega_\infty.$$

Hence

$$\mathbf{T}_1\mathbf{f} = \{T_{1,r} f_r : r \in \Omega_\infty\} = \{g_r : r \in \Omega_\infty\},$$

because

$$\sum_{r=1}^{\infty} \|T_{1,r} f_r\|_r^2 = \sum_{r=1}^{\infty} \|g_r\|_r^2 = \|\mathbf{g}\|_{\mathbf{H}}^2 < \infty.$$

Therefore, as required,

$$\mathbf{f} \in \mathbf{D}(\mathbf{T}_1) \text{ and also } \mathbf{T}_1\mathbf{f} = \mathbf{g}.$$

A similar argument proves that $\mathbf{T}_0$ on $\mathbf{D}(\mathbf{T}_0)$ is a closed operator in $\mathbf{H}$.

Next, we demonstrate that $\mathbf{D}(\mathbf{T}_0)$ is dense in the Hilbert space $\mathbf{H}$. Clearly $\mathbf{D}(\mathbf{T}_{00}) \subseteq \mathbf{D}(\mathbf{T}_0)$ since, see Lemma 5.2 and Theorem 5.1 below,

$$\mathbf{f} \in \mathbf{D}(\mathbf{T}_{00}) \Longrightarrow [f_r : v_r]_r = 0 \text{ for all } r \in \Omega \text{ and all } \mathbf{v} = \{v_r : r \in \Omega\} \in \mathbf{D}(\mathbf{T}_1),$$

and so it is only necessary to show that $\mathbf{D}(\mathbf{T}_{00})$ is dense in $\mathbf{H}$.

Now take any vector $\mathbf{h} = \{h_r : r \in \Omega\} \in \mathbf{H}$ and we seek to approximate $\mathbf{h}$ by vectors in $\mathbf{D}(\mathbf{T}_{00})$. We can assume, for our purposes and without loss of generality, that only a finite number of the components h_r are non-zero - say $h_r \in L_r^2$ for $r = 1, 2, \ldots, l$.

It is known, see [**9**], that the manifold

$$D_0(T_{1,r}) := \{f_r \in D(T_{1,r}) : \text{ supp}(f_r) \text{ is a compact subset interior to } I_r\}$$

is dense in L_r^2, for each $r \in \Omega$. Hence, for each prescribed number $\varepsilon > 0$, there exist functions $\{f_r \in D_0(T_{1,r}) : r = 1, 2, \ldots, l\}$ such that

$$\sum_{r=1}^{l} \|f_r - h_r\|_r^2 < \varepsilon^2.$$

Now define the vector $\mathbf{f} \in \mathbf{H}$ by the components $\{f_r : r = 1, 2, \ldots, l\}$ and all other components $f_r = 0$. Then certainly $\mathbf{f} \in \mathbf{D}(\mathbf{T}_{00})$ and $\|\mathbf{f} - \mathbf{h}\|_{\mathbf{H}} < \varepsilon$, as required. □

Definition 5.3. Let $\{I_r, M_r, w_r : r \in \Omega\}$ be a multi-interval system with corresponding Hilbert space $\mathbf{H} = \sum_{r\in\Omega} \oplus L_r^2$, and minimal and maximal operators

$$\mathbf{T}_0 \subseteq \mathbf{T}_1 \text{ on } \mathbf{D}(\mathbf{T}_0) \subseteq \mathbf{D}(\mathbf{T}_1) \subset \mathbf{H}, \tag{5.13}$$

as above. Then define the *boundary form* $[\cdot : \cdot]_{\mathbf{D}} : \mathbf{D}(\mathbf{T}_1) \times \mathbf{D}(\mathbf{T}_1) \to \mathbb{C}$ by

$$[\mathbf{f} : \mathbf{g}]_{\mathbf{D}} := \langle \mathbf{T}_1\mathbf{f}, \mathbf{g}\rangle_{\mathbf{H}} - \langle \mathbf{f}, \mathbf{T}_1\mathbf{g}\rangle_{\mathbf{H}} \text{ for all } \mathbf{f}, \mathbf{g} \in \mathbf{D}(\mathbf{T}_1). \tag{5.14}$$

Clearly $[\cdot : \cdot]$ is a skew-hermitian sesquilinear (or conjugate-bilinear) complex-valued form on $\mathbf{D}(\mathbf{T}_1)$.

The next Lemma 5.2 and Theorem 5.1 are proved in [**15**, Theorem 2.1] for the case of countable Ω, but the same proof holds in the general case - as indicated in Lemma 5.1.

LEMMA 5.2. *Let $\{I_r, M_r, w_r : r \in \Omega\}$ be a multi-interval system with minimal and maximal operators $\mathbf{T}_0$ on $\mathbf{D}(\mathbf{T}_0)$ and $\mathbf{T}_1$ on $\mathbf{D}(\mathbf{T}_1)$, respectively, on the Hilbert space $\mathbf{H} = \sum_{r\in\Omega} \oplus L_r^2$, as before.*

Then for $\mathbf{f} = \{f_r : r \in \Omega\}$ and $\mathbf{g} = \{g_r : r \in \Omega\}$ in $\mathbf{D}(\mathbf{T}_1)$,

$$[\mathbf{f} : \mathbf{g}]_{\mathbf{D}} = \sum_{r\in\Omega} [f_r : g_r]_r \tag{5.15}$$

where the absolute convergence of the series is assured from

$$|[\mathbf{f} : \mathbf{g}]_{\mathbf{D}}| \leq \sum_{r\in\Omega} |[f_r : g_r]_r| \leq \|\mathbf{g}\|_{\mathbf{H}} \|\mathbf{T}_1\mathbf{f}\|_{\mathbf{H}} + \|\mathbf{f}\|_{\mathbf{H}} \|\mathbf{T}_1\mathbf{g}\|_{\mathbf{H}}.$$

From this Lemma it follows that $[\cdot : \cdot]_{\mathbf{D}}$ is a continuous map of the product space $\mathbf{D}(\mathbf{T}_1) \times \mathbf{D}(\mathbf{T}_1) \to \mathbb{C}$ in terms of the $\mathbf{T}_1$-operator or graph $\mathbf{D}$-norm and metric on $\mathbf{D}(\mathbf{T}_1)$ (see the definition of the Hilbert space $\mathbf{D}(\mathbf{T}_1)$ given in (6.11) below), but not continuous (even in one variable) in the $\mathbf{H}$-metric.

THEOREM 5.1. *Let* $\{I_r, M_r, w_r : r \in \Omega\}$ *be a multi-interval system with minimal and maximal operators* $\mathbf{T}_0$ *on* $\mathbf{D}(\mathbf{T}_0)$ *and* $\mathbf{T}_1$ *on* $\mathbf{D}(\mathbf{T}_1)$, *respectively, on the Hilbert space* $\mathbf{H} = \sum_{r\in\Omega} \oplus L_r^2$, *as before.*

Then the operators

$$\mathbf{T}_0 \subseteq \mathbf{T}_1 \text{ on } \mathbf{D}(\mathbf{T}_0) \subseteq \mathbf{D}(\mathbf{T}_1) \subset \mathbf{H} \tag{5.16}$$

satisfy the adjoint relations (with respect to $\mathbf{H}$*)*

$$\mathbf{T}_0^* = \mathbf{T}_1 \text{ and } \mathbf{T}_1^* = \mathbf{T}_0. \tag{5.17}$$

Furthermore, for each $\mathbf{f} \in \mathbf{D}(\mathbf{T}_1)$

$$[\mathbf{f} : \mathbf{D}(\mathbf{T}_1)]_{\mathbf{D}} = 0 \text{ if and only if } \mathbf{f} \in \mathbf{D}(\mathbf{T}_0). \tag{5.18}$$

Accordingly, $\mathbf{T}_0$ *on* $\mathbf{D}(\mathbf{T}_0)$ *is a closed symmetric operator, and* $\mathbf{T}_1$ *on* $\mathbf{D}(\mathbf{T}_1)$ *is also closed - and both* $\mathbf{D}(\mathbf{T}_0)$ *and* $\mathbf{D}(\mathbf{T}_1)$ *are dense in* $\mathbf{H}$.

Moreover $\mathbf{T}_0$ *on* $\mathbf{D}(\mathbf{T}_0)$ *is the closure of the symmetric operator* $\mathbf{T}_{00}$ *on* $\mathbf{D}(\mathbf{T}_{00}) := \mathbf{D}_0(\mathbf{T}_1)$; *that is*

$$\overline{\mathbf{T}}_{00} = \mathbf{T}_0 \text{ and so } \mathbf{T}_{00}^* = \mathbf{T}_1. \tag{5.19}$$

Hence

$$\mathbf{T}_{00} \subseteq \mathbf{T}_0 \subseteq \mathbf{T}_1 \text{ on } \mathbf{D}(\mathbf{T}_{00}) \subseteq \mathbf{D}(\mathbf{T}_0) \subseteq \mathbf{D}(\mathbf{T}_1). \tag{5.20}$$

PROOF. The conclusions (5.16), (5.17) and (5.18) are established in [**15**] (when Ω is countable, but this restriction does not effect the generality of the result - as mentioned earlier) where they are proved from the corresponding results valid for each $r \in \Omega$ - compare with (2.21) to (2.27) of Section 2 above. Moreover (5.18) and Lemma 5.1 also show that $\mathbf{T}_{00} \subseteq \mathbf{T}_0$ are both symmetric operators in $\mathbf{H}$.

We consequently discuss only the result (5.19) involving the operator $\mathbf{T}_{00}$ on $\mathbf{D}(\mathbf{T}_{00}) := \mathbf{D}_0(\mathbf{T}_1)$, as defined in Lemma 5.1, and its closure $\overline{\mathbf{T}}_{00}$ and adjoint $\mathbf{T}_{00}^*$. We first proceed to demonstrate that

$$\overline{\mathbf{T}}_{00} = \mathbf{T}_0 \text{ on } \mathbf{D}(\overline{\mathbf{T}}_{00}) = \mathbf{D}(\mathbf{T}_0),$$

noting that both $\mathbf{T}_{00}$ and $\mathbf{T}_0$ are restrictions of $\mathbf{T}_1$.

Recall that a vector $\mathbf{g} \in \mathbf{H}$ belongs to the domain $\mathbf{D}(\overline{\mathbf{T}}_{00})$ just in case there exists a sequence $\{\mathbf{f}_k \in \mathbf{D}(\mathbf{T}_{00}) : k \in \mathbb{N}\}$ such that, in $\mathbf{H}$,

$$\lim_{k\to\infty} \mathbf{f}_k = \mathbf{g} \text{ and } \lim_{k\to\infty} \mathbf{T}_1\mathbf{f}_k = \mathbf{h},$$

and then we define $\overline{\mathbf{T}}_{00}\mathbf{g} := \mathbf{h}$.

Since $\mathbf{D}(\mathbf{T}_{00}) \subseteq \mathbf{D}(\mathbf{T}_0)$ and since $\mathbf{T}_0$ is a closed operator, it is evident that $\mathbf{D}(\overline{\mathbf{T}}_{00}) \subseteq \mathbf{D}(\mathbf{T}_0)$ and $\overline{\mathbf{T}}_{00}$ is a restriction of $\mathbf{T}_0$ (and hence a restriction of $\mathbf{T}_1$).

We seek the reverse inclusion, that $\mathbf{D}(\mathbf{T}_0) \subseteq \mathbf{D}(\overline{\mathbf{T}}_{00})$. For this purpose take any vector $\mathbf{g} \in \mathbf{D}(\mathbf{T}_0)$; then we must verify that $\mathbf{g}$ can be arbitrarily closely approximated (in the sense of the norm in $\mathbf{H}$) by functions $\mathbf{f} \in \mathbf{D}(\mathbf{T}_{00})$, with $\mathbf{T}_1\mathbf{f}$ simultaneously approximating $\mathbf{T}_0\mathbf{g} = \mathbf{T}_1\mathbf{g}$. However, since we are dealing with approximations, we can assume, without loss of generality, that we select $\mathbf{g} \in \mathbf{D}(\mathbf{T}_0)$ so that only a finite number of its components $\{g_r\}$ are non-zero; that is, there exists some integer $N \geq 1$ such that $g_r = 0$ for all $r > N$ and hence

$$(\mathbf{T}_1\mathbf{g})_r = T_{1,r}g_r = 0 \text{ for all } r > N$$

(using some convenient ordering and numbering of the elements of $\{g_r\}$).

Now fix a positive number $\varepsilon > 0$, and take functions $f_r \in D_0(T_{1,r})$ for each $r = 1, 2, \ldots, N$, with

$$\|f_r - g_r\|_r^2 < \frac{\varepsilon^2}{2N} \text{ and } \|T_{1,r}f_r - T_{1,r}g_r\|_r^2 < \frac{\varepsilon^2}{2N}.$$

This is possible because, for each fixed $r \leq N$, we have $f_r \in D(T_{00,r})$ and $g_r \in D(T_{0,r})$ with

$$\overline{T}_{00,r} = T_{0,r} \text{ on } D(\overline{T}_{00,r}) = D(T_{0,r}) \text{ in } L_r^2;$$

see [**9**]. Now define $\mathbf{f} \in \mathbf{D}(\mathbf{T}_{00})$ with already specified components $\{f_r : r = 1, 2, \ldots, N\}$, and all other components zero. In this case

$$\|\mathbf{f} - \mathbf{g}\|_{\mathbf{H}}^2 \leq \varepsilon^2/2 \text{ and } \|\mathbf{T}_1\mathbf{f} - \mathbf{T}_1\mathbf{g}\|_{\mathbf{H}}^2 \leq \varepsilon^2/2.$$

We can now repeat the process involved in the construction of $\mathbf{f}$, for each choice of $\varepsilon = k^{-1}$ for all $k \in \mathbb{N}$, and so obtain the desired sequence $\{\mathbf{f}_k : k \in \mathbb{N}\}$, with limits

$$\lim_{k\to\infty} \mathbf{f}_k = \mathbf{g} \text{ and } \lim_{k\to\infty} \mathbf{T}_1\mathbf{f}_k = \mathbf{T}_1\mathbf{g} \text{ in } \mathbf{H}.$$

Hence, each $\mathbf{g} \in \mathbf{D}(\mathbf{T}_0)$ belongs to $\mathbf{D}(\overline{\mathbf{T}}_{00})$, and furthermore $\overline{\mathbf{T}}_{00}\mathbf{g} = \mathbf{T}_0\mathbf{g}$. Therefore

$$\mathbf{D}(\overline{\mathbf{T}}_{00}) = \mathbf{D}(\mathbf{T}_0) \text{ with } \overline{\mathbf{T}}_{00}\mathbf{g} = \mathbf{T}_0\mathbf{g},$$

and we have now demonstrated that $\mathbf{T}_0$ is the minimal closed operator extending $\mathbf{T}_{00}$ in $\mathbf{H}$; that is $\overline{\mathbf{T}}_{00} = \mathbf{T}_0$. From classical results in the theory of Hilbert spaces, see [**4**, Chapter XII.4.8 and 7.1] and [**26**], we then know that

$$\mathbf{T}_{00}^* = (\overline{\mathbf{T}}_{00})^* = \mathbf{T}_0^* = \mathbf{T}_1.$$

This completes the proof of Theorem 5.1. □

Remark 5.2. As usual, see [4, Chapter XII], an operator $\mathbf{T}$ on $\mathbf{D}(\mathbf{T}) \subset \mathbf{H}$ is self-adjoint in case $\mathbf{T}^* = \mathbf{T}$ on $\mathbf{D}(\mathbf{T}^*) = \mathbf{D}(\mathbf{T})$; thus $\mathbf{T}_0 = \mathbf{T}_1$ if and only if $\mathbf{T}_0$ is self-adjoint, compare (2.27) and (2.28).

We can now define the boundary space $\mathsf{S} := \mathbf{D}(\mathbf{T}_1)/\mathbf{D}(\mathbf{T}_0)$ as a complex symplectic space - possibly of infinite dimensions.

Definition 5.4. Let $\{I_r, M_r, w_r : r \in \Omega\}$ be a multi-interval system, with minimal and maximal operators $\mathbf{T}_0$ on $\mathbf{D}(\mathbf{T}_0)$ and $\mathbf{T}_1$ on $\mathbf{D}(\mathbf{T}_1)$, respectively, in the Hilbert space $\mathbf{H} = \sum_{r\in\Omega} \oplus L_r^2$, with the corresponding boundary form $[\cdot : \cdot]_{\mathbf{D}} : \mathbf{D}(\mathbf{T}_1) \times \mathbf{D}(\mathbf{T}_1) \to \mathbb{C}$, defined in (5.14),

Then define the boundary complex symplectic space

$$\mathsf{S} := \mathbf{D}(\mathbf{T}_1)/\mathbf{D}(\mathbf{T}_0) \text{ with } [\mathsf{f} : \mathsf{g}]_{\mathsf{S}} := [\mathbf{f} : \mathbf{g}]_{\mathbf{D}} \tag{5.21}$$

where $\mathsf{f} := \Psi\mathbf{f} = \{\mathbf{f} + \mathbf{D}(\mathbf{T}_0)\}$, $\mathsf{g} := \Psi\mathbf{g} = \{\mathbf{g} + \mathbf{D}(\mathbf{T}_0)\}$ are cosets belonging to the identification or quotient space S.

The natural projection map of $\mathbf{D}(\mathbf{T}_1)$ onto S

$$\Psi : \mathbf{D}(\mathbf{T}_1) \to \mathsf{S} \text{ and } \mathbf{f} \to \mathsf{f} = \{\mathbf{f} + \mathbf{D}(\mathbf{T}_0)\} \tag{5.22}$$

is a linear surjection, and we also denote the corresponding set-map by Ψ, and the inverse set-map by Ψ^{-1} whenever convenient.

6. Boundary symplectic spaces for multi-interval systems

Throughout this section a multi-interval quasi-differential system, as first introduced in (1.1) and presented fully in Definition 2.1, is denoted by

$$\{I_r, M_r, w_r : r \in \Omega\}. \tag{6.1}$$

Such a multi-interval system defines linear operators in the system Hilbert space $\mathbf{H} = \sum_{r\in\Omega} \oplus L_r^2$, consisting of vectors

$$\mathbf{f} = \{f_r \in L_r^2 : r \in \Omega\} \text{ with } \|\mathbf{f}\|_{\mathbf{H}}^2 = \sum_{r\in\Omega} \|f_r\|_r^2 < +\infty, \tag{6.2}$$

and with the corresponding scalar product

$$\langle \mathbf{f}, \mathbf{g}\rangle_{\mathbf{H}} = \sum_{r\in\Omega} \langle f_r, g_r\rangle_r\,, \tag{6.3}$$

as explained in Definition 5.1 and (5.1) to (5.7), above.

In particular, the minimal operator $\mathbf{T}_0$ on the dense domain $\mathbf{D}(\mathbf{T}_0) \subset \mathbf{H}$ is the minimal closed symmetric operator (5.10), (5.11) generated by the multi-interval system (6.1). Also the maximal operator $\mathbf{T}_1$ on $\mathbf{D}(\mathbf{T}_1) \subset \mathbf{H}$ is the adjoint, see Theorem 5.1,

$$\mathbf{T}_1 = \mathbf{T}_0^* \text{ so that } \mathbf{T}_0 = \mathbf{T}_1^* \text{ on } \mathbf{D}(\mathbf{T}_0) \subseteq \mathbf{D}(\mathbf{T}_1). \tag{6.4}$$

In terms of these two operators we define (see Definition 5.4) the system boundary space, as the quotient vector space,

$$\mathsf{S} := \mathbf{D}(\mathbf{T}_1)/\mathbf{D}(\mathbf{T}_0) \tag{6.5}$$

of cosets $\mathsf{f} := \{\mathbf{f} + \mathbf{D}(\mathbf{T}_0)\}$, which is a complex symplectic space with the symplectic product

$$[\mathsf{f} : \mathsf{g}]_{\mathsf{S}} := [\mathbf{f} : \mathbf{g}]_{\mathbf{D}} = [\mathbf{f} + \mathbf{D}(\mathbf{T}_0) : \mathbf{g} + \mathbf{D}(\mathbf{T}_0)]_{\mathbf{D}} = \sum_{r\in\Omega} [f_r : g_r]_r \tag{6.6}$$

in terms of the symplectic product on $\mathbf{D}(\mathbf{T}_1)$, see (5.14) and (5.15),

$$[\mathbf{f} : \mathbf{g}]_{\mathbf{D}} := \langle \mathbf{T}_1\mathbf{f}, \mathbf{g}\rangle_{\mathbf{H}} - \langle \mathbf{f}, \mathbf{T}_1\mathbf{g}\rangle_{\mathbf{H}} \text{ for all } \mathbf{f}, \mathbf{g} \in \mathbf{D}(\mathbf{T}_1). \tag{6.7}$$

Further, in this section, we define a norm $\|\cdot\|_{\mathsf{S}}$ on S to fix S as a complete Hilbert space. We then use all these structures on the boundary space S to describe and classify all self-adjoint extensions $\mathbf{T}$ of $\mathbf{T}_0$ through a generalization of the classical theorem of Glazman-Krein-Naimark. For the Everitt-Markus version of the GKN-Theorem, see [**8**] and [**10**].

For each self-adjoint extension $\mathbf{T}$ on $\mathbf{D}(\mathbf{T})$ of $\mathbf{T}_0$ on $\mathbf{D}(\mathbf{T}_0)$, it is clear that the following bounds are satisfied

$$\mathbf{T}_0 \subseteq \mathbf{T} \subseteq \mathbf{T}_1 \text{ on } \mathbf{D}(\mathbf{T}_0) \subseteq \mathbf{D}(\mathbf{T}) \subseteq \mathbf{D}(\mathbf{T}_1). \tag{6.8}$$

Moreover, according to the Stone-von Neumann result on the existence of the direct sum decomposition of the linear space $\mathbf{D}(\mathbf{T}_1)$ into three linear submanifolds, see [**4**, Chapter XII.4.10],

$$\mathbf{D}(\mathbf{T}_1) = \mathbf{D}(\mathbf{T}_0) \dotplus \mathbf{N}^- \dotplus \mathbf{N}^+ \tag{6.9}$$

so that each function $\mathbf{f} \in \mathbf{D}(\mathbf{T}_1)$ has the unique representation

$$\mathbf{f} = \mathbf{f}_0 + \mathbf{f}_- + \mathbf{f}_+ \text{ with } \mathbf{f}_0 \in \mathbf{D}(\mathbf{T}_0) \text{ and } \mathbf{f}_\pm \in \mathbf{N}^\pm$$

respectively. Here the deficiency spaces $\mathbf{N}^{\pm}$ are defined, as usual, by

$$\mathbf{N}^{\pm} := \{\mathbf{f} \in \mathbf{D}(\mathbf{T}_1) : \mathbf{T}_1\mathbf{f} = \pm i f\}. \tag{6.10}$$

Now we reformulate these results using the basic structures of $\mathbf{D}(\mathbf{T}_1)$:

(1) $\mathbf{D}(\mathbf{T}_1)$ is a complex Hilbert space, with the $\mathbf{T}_1$-operator or graph norm $\|\cdot\|_{\mathbf{D}}$, corresponding to the defined scalar product

$$\langle \mathbf{f}, \mathbf{g} \rangle_{\mathbf{D}} := \langle \mathbf{f}, \mathbf{g} \rangle_{\mathbf{H}} + \langle \mathbf{T}_1\mathbf{f}, \mathbf{T}_1\mathbf{g} \rangle_{\mathbf{H}} \text{ so that } \|\mathbf{f}\|_{\mathbf{D}}^{\mathbf{2}} = \langle \mathbf{f}, \mathbf{f} \rangle_{\mathbf{D}}. \tag{6.11}$$

(2) $\mathbf{D}(\mathbf{T}_1)$ has a (degenerate) symplectic product

$$[\mathbf{f} : \mathbf{g}]_{\mathbf{D}} := \langle \mathbf{T}_1\mathbf{f}, \mathbf{g} \rangle_{\mathbf{H}} - \langle \mathbf{f}, \mathbf{T}_1\mathbf{g} \rangle_{\mathbf{H}} \text{ for all } \mathbf{f}, \mathbf{g} \in \mathbf{D}(\mathbf{T}_1). \tag{6.12}$$

The next theorem recapitulates and summarizes many of the results of Section 5, in view of the Hilbert space and symplectic structures on $\mathbf{D}(\mathbf{T}_1)$.

THEOREM 6.1. *Consider the Stone-von Neumann direct sum decomposition*

$$\mathbf{D}(\mathbf{T}_1) = \mathbf{D}(\mathbf{T}_0) \oplus \mathbf{N}^- \oplus \mathbf{N}^+ \textit{ with } \mathbf{f} = \mathbf{f}_0 + \mathbf{f}_- + \mathbf{f}_+ \tag{6.13}$$

respectively, as before. Then the following conclusions hold and account for the special notation as in (6.13):

(1) *The three summands in* (6.13) *are pairwise orthogonal, closed Hilbert subspaces of the Hilbert space* $\mathbf{D}(\mathbf{T}_1)$; *that is*

$$\langle \mathbf{N}^-, \mathbf{N}^+ \rangle_{\mathbf{D}} = 0 \textit{ and } \langle \mathbf{N}^{\pm}, \mathbf{D}(\mathbf{T}_0) \rangle_{\mathbf{D}} = 0. \tag{6.14}$$

In addition $\mathbf{N}^{\pm}$ *are each closed Hilbert subspaces of* $\mathbf{H}$; *in fact, the two norms agree on each* $\mathbf{N}^{\pm}$ *(excepting for a factor* $\sqrt{2}$), *since*

$$\|\mathbf{f}_-\|_{\mathbf{D}}^2 = 2\,\|\mathbf{f}_-\|_{\mathbf{H}}^2 \textit{ and } \|\mathbf{f}_+\|_{\mathbf{D}}^2 = 2\,\|\mathbf{f}_+\|_{\mathbf{H}}^2. \tag{6.15}$$

We define the deficiency indices $\mathbf{d}^-$ *and* $\mathbf{d}^+$ *by*

$$\mathbf{d}^{\pm} := \dim(\mathbf{N}^{\pm}) \tag{6.16}$$

invoking the cardinality of complete orthogonal bases for the respective Hilbert spaces.

(2) *The three summands are pairwise symplectically orthogonal in* $\mathbf{D}(\mathbf{T}_1)$, *that is*

$$[\mathbf{N}^- : \mathbf{N}^+]_{\mathbf{D}} = 0 \textit{ and } [\mathbf{N}^{\pm} : \mathbf{D}(\mathbf{T}_0)]_{\mathbf{D}} = 0. \tag{6.17}$$

(3)

$$\mathbf{D}(\mathbf{T}_0) = \{\mathbf{f} \in \mathbf{D}(\mathbf{T}_1) : [\mathbf{f} : \mathbf{D}(\mathbf{T}_1)]_{\mathbf{D}} = 0\}. \tag{6.18}$$

(4) *The complex linear space* $\mathbf{N}^- \oplus \mathbf{N}^+$ *is a Hilbert space under the norm* $\|\cdot\|_{\mathbf{D}}$; *moreover*

$$\langle \mathbf{f}_- + \mathbf{f}_+, \mathbf{N}^- \rangle_{\mathbf{D}} = 0 \textit{ implies that } \mathbf{f}_- = 0; \tag{6.19}$$

thus $\mathbf{N}^+$ *is the orthogonal complement of* $\mathbf{N}^-$ *in* $\mathbf{N}^- \oplus \mathbf{N}^+$, *and similarly* $\mathbf{N}^-$ *is the orthogonal complement of* $\mathbf{N}^+$.

(5) *The complex vector space* $\mathbf{N}^- \oplus \mathbf{N}^+$ *is a complex symplectic space with the symplectic product (see* 6.12)

$$\begin{aligned} [\mathbf{f}_- + \mathbf{f}_+ : \mathbf{g}_- + \mathbf{g}_+]_{\mathbf{D}} &= -2i\,\langle \mathbf{f}_-, \mathbf{g}_- \rangle_{\mathbf{H}} + 2i\,\langle \mathbf{f}_+, \mathbf{g}_+ \rangle_{\mathbf{H}} \\ &= [\mathbf{f}_- : \mathbf{g}_-]_{\mathbf{D}} + [\mathbf{f}_+ : \mathbf{g}_+]_{\mathbf{D}}. \end{aligned} \tag{6.20}$$

Moreover

(6.21) $$[\mathbf{f}_- + \mathbf{f}_+ : \mathbf{N}^-]_{\mathbf{D}} = 0 \text{ implies that } \mathbf{f}_- = 0$$

so $\mathbf{N}^-$ *and* $\mathbf{N}^+$ *are symplectic orthocomplements in* $\mathbf{N}^- \oplus \mathbf{N}^+$.

(6) *The boundary space*

(6.22) $$\mathsf{S} := \mathbf{D}(\mathbf{T}_1)/\mathbf{D}(\mathbf{T}_0) \text{ of cosets } \mathsf{f} = \{\mathbf{f} + \mathbf{D}(\mathbf{T}_0)\},$$

is a complex symplectic space with the symplectic product

(6.23) $$[\mathsf{f} : \mathsf{g}]_{\mathsf{S}} := [\mathbf{f} + \mathbf{D}(\mathbf{T}_0) : \mathbf{g} + \mathbf{D}(\mathbf{T}_0)]_{\mathbf{D}} = [\mathbf{f} : \mathbf{g}]_{\mathbf{D}}$$

for $\mathsf{f} = \{\mathbf{f} + \mathbf{D}(\mathbf{T}_0)\}, \mathsf{g} = \{\mathbf{g} + \mathbf{D}(\mathbf{T}_0)\}$ *with* $\mathbf{f}, \mathbf{g} \in \mathbf{D}(\mathbf{T}_1)$.

(7) *There is a natural symplectic isomorphism of* $\mathbf{N}^- \oplus \mathbf{N}^+$ *onto* S, *which is denoted by* $\mathsf{S} \approx \mathbf{N}^- \oplus \mathbf{N}^+$, *namely the natural projection map* Ψ *of* (5.22) *restricted to* $\mathbf{N}^- \oplus \mathbf{N}^+$,

(6.24) $$\Psi : \mathbf{N}^- \oplus \mathbf{N}^+ \to \mathsf{S} \qquad \mathbf{f}_- + \mathbf{f}_+ \to \mathsf{f}$$

where $\mathsf{f} = \Psi(\mathbf{f}_- + \mathbf{f}_+) = \{\mathbf{f}_- + \mathbf{f}_+ + \mathbf{D}(\mathbf{T}_0)\}$.

PROOF. The conclusion 1 is a classical result in Hilbert space operator theory, see [**4**, Chapter XII.4]. Then conclusions 2 through 7 are straightforward, direct calculations following the definitions and assertions in Section 5 above. □

Note The metric on the Hilbert space $\mathbf{N}^- \oplus \mathbf{N}^+$ is defined by the $\mathbf{D}$-norm $\|\cdot\|_{\mathbf{D}}$, wherein the subspaces $\mathbf{N}^-$ and $\mathbf{N}^+$ are orthocomplements. However $\|\mathbf{f}_-\|_{\mathbf{D}} = \sqrt{2}\,\|\mathbf{f}_-\|_{\mathbf{H}}$ for all $\mathbf{f}_- \in \mathbf{N}^-$ (and similarly for $\mathbf{f}_+ \in \mathbf{N}^+$), so a computation shows

(6.25) $$\|\mathbf{f}_- + \mathbf{f}_+\|_{\mathbf{D}}^2 = \|\mathbf{f}_-\|_{\mathbf{D}}^2 + \|\mathbf{f}_+\|_{\mathbf{D}}^2 = 2\,\|\mathbf{f}_-\|_{\mathbf{H}}^2 + 2\,\|\mathbf{f}_+\|_{\mathbf{H}}^2$$

and then

(6.26) $$\|\mathbf{f}_- + \mathbf{f}_+\|_{\mathbf{H}} \le \tfrac{1}{\sqrt{2}}\left(\|\mathbf{f}_-\|_{\mathbf{D}} + \|\mathbf{f}_+\|_{\mathbf{D}}\right).$$

Definition 6.1. Consider the boundary complex symplectic space

$$\mathsf{S} = \mathbf{D}(\mathbf{T}_1)/\mathbf{D}(\mathbf{T}_0) \text{ with symplectic form } [\cdot : \cdot]_{\mathsf{S}}$$

and the deficiency complex symplectic space

$$\mathbf{N}^- \oplus \mathbf{N}^+ \text{ with symplectic form } [\cdot : \cdot]_{\mathbf{D}}.$$

By Theorem 6.1, (6.24), there is a natural (bijective) symplectic isomorphism of $\mathbf{N}^- \oplus \mathbf{N}^+$ onto S

(6.27) $$\Psi : \mathbf{N}^- \oplus \mathbf{N}^+ \to \mathsf{S} \text{ with } \mathbf{f}_- + \mathbf{f}_+ \to \ \mathsf{f} = \{\mathbf{f}_- + \mathbf{f}_+ + \mathbf{D}(\mathbf{T}_0)\}$$

where $\mathbf{f} = \mathbf{f}_- + \mathbf{f}_+$ defines the coset $\mathsf{f} \in \mathsf{S}$. Since $\mathbf{N}^- \oplus \mathbf{N}^+$ is a Hilbert space under the norm $\|\cdot\|_{\mathbf{D}}$, the map Ψ may be used to carry this norm to S so defining the norm $\|\cdot\|_{\mathsf{S}}$ and thus making S into a complex Hilbert space. Namely, for each $\mathsf{f} = \{\mathbf{f}_- + \mathbf{f}_+ + \mathbf{D}(\mathbf{T}_0)\} \in \mathsf{S}$ define, using (6.25) and (6.26),

(6.28) $$\|\mathsf{f}\|_{\mathsf{S}}^2 := \|\mathbf{f}_- + \mathbf{f}_+\|_{\mathbf{D}}^2 = \|\mathbf{f}_-\|_{\mathbf{D}}^2 + \|\mathbf{f}_+\|_{\mathbf{D}}^2 = 2\,\|\mathbf{f}_-\|_{\mathbf{H}}^2 + 2\,\|\mathbf{f}_+\|_{\mathbf{H}}^2.$$

Further, the scalar product in S is given by

(6.29) $$\langle \mathsf{f}, \mathsf{g}\rangle_{\mathsf{S}} := \langle \mathbf{f}_- + \mathbf{f}_+, \mathbf{g}_- + \mathbf{g}_+\rangle_{\mathbf{D}} = \langle \mathbf{f}_-, \mathbf{g}_-\rangle_{\mathbf{D}} + \langle \mathbf{f}_+, \mathbf{g}_+\rangle_{\mathbf{D}}.$$

Remark 6.1. The concepts and results in Theorem 6.1 and Definition 6.1 also hold for single-interval systems where $\Omega = \{1\}$. In particular if we now fix one value of r

in a multi-interval system $\{I_r, M_r, w_r : r \in \Omega\}$, then the system Hilbert space is L_r^2, with minimal and maximal operators $T_{0,r}$ on $D(T_{0,r})$ and $T_{1,r}$ on $D(T_{1,r}) \subset L_r^2$, as in Section 2. Following the notations of (6.11) and (6.12), we assert that $D(T_{1,r})$ is a Hilbert space with the $T_{1,r}$-operator norm $\|\cdot\|_{D_r}$ and also it bears the symplectic product $[\cdot:\cdot]_{D_r} \equiv [\cdot:\cdot]_r$.

Again, the boundary complex symplectic space

$$\mathsf{S}_r := D(T_{1,r})/D(T_{0,r}) \text{ with } [\cdot:\cdot]_r,$$

is symplectically isomorphic to $N_r^- \oplus N_r^+$ and this is used to define the Hilbert space norm $\|\cdot\|_{\mathsf{S}_r}$ on S_r, as in Definition 6.1.

Here S_r is a finite dimensional Hilbert space, but for general multi-interval systems (6.1) the boundary space S is an infinite dimensional Hilbert space and we must distinguish between linear submanifolds of S and closed subspaces of S, as illustrated in the next Lemma.

LEMMA 6.1. *Let $\{I_r, M_r, w_r : r \in \Omega\}$ be a multi-interval system with the boundary complex symplectic space $\mathsf{S} = \mathbf{D}(\mathbf{T}_1)/\mathbf{D}(\mathbf{T}_0)$, which is a Hilbert space with the norm $\|\cdot\|_{\mathsf{S}}$ as in Definition* 6.1.

Then a complete Lagrangian submanifold $\mathsf{L} \subset \mathsf{S}$ is necessarily a closed Hilbert subspace of S.

PROOF. Since S is isomorphic with $\mathbf{N}^- \oplus \mathbf{N}^+$ both as a complex symplectic space and as a Hilbert space, it can be assumed that L is a complete Lagrangian manifold in $\mathbf{N}^- \oplus \mathbf{N}^+$, without loss of generality.

Take any vector $\mathbf{f} = \mathbf{f}_- + \mathbf{f}_+ \in \overline{\mathsf{L}}$, the closure of L in the Hilbert space $\mathbf{N}^- \oplus \mathbf{N}^+$. Then there exists a sequence of vectors $\{\mathbf{v}_k \in \mathsf{L} : k \in \mathbb{N}\}$ with $\lim_{k\to\infty} \mathbf{v}_k = \mathbf{f}$, in the sense of the $\mathbf{D}$-norm metric in $\mathbf{N}^- \oplus \mathbf{N}^+$.

Now fix any vector $\mathbf{v} \in \mathsf{L}$ and note that $[\mathbf{v}_k : \mathbf{v}]_{\mathbf{D}} = 0$, since L is a Lagrangian manifold. Now the map

$$\mathbf{N}^- \oplus \mathbf{N}^+ \to \mathbb{C} \ i.e.\ \mathbf{u} \to [\mathbf{u} : \mathbf{v}]_{\mathbf{D}}$$

is continuous for $\mathbf{u} \in \mathbf{N}^- \oplus \mathbf{N}^+$ (in terms of the $\mathbf{D}$-norm on $\mathbf{N}^- \oplus \mathbf{N}^+$, see Lemma 5.2), and hence $[\mathbf{f} : \mathbf{v}] = 0$ for each fixed $\mathbf{v} \in \mathsf{L}$. However, $[\mathbf{f} : \mathsf{L}] = 0$ implies that $\mathbf{f} \in \mathsf{L}$, since L is assumed to be a complete Lagrangian.

Thus $\overline{\mathsf{L}} \subseteq \mathsf{L}$ and so L is a closed subspace of $\mathbf{N}^- \oplus \mathbf{N}^+$, as required. □

Remark 6.2. The conclusion of Lemma 6.1 is clearly valid if we merely assume that S has a topology for which the maps

$$\mathsf{S} \to \mathbb{C},\ \mathsf{u} \to [\mathsf{u} : \mathsf{v}]_{\mathsf{S}} \tag{6.30}$$

are continuous, for each fixed $\mathsf{v} \in \mathsf{S}$. The weakest topology satisfying this requirement is called the "symplectic weak topology" on S, and has been seriously investigated in [**11**]. It is evident that a subset of S which is closed in this symplectic weak topology is necessarily closed in the metric topology of the Hilbert space S.

It is now possible to state a new version of the Glazman-Krein-Naimark Theorem concerning self-adjoint extensions of the minimal operator $\mathbf{T}_0$ on $\mathbf{D}(\mathbf{T}_0) \subset \mathbf{H}$.

THEOREM 6.2. *Let $\{I_r, M_r, w_r : r \in \Omega\}$ be a multi-interval system, with minimal and maximal operators $\mathbf{T}_0$ on $\mathbf{D}(\mathbf{T}_0)$ and $\mathbf{T}_1$ on $\mathbf{D}(\mathbf{T}_1)$, respectively in the system Hilbert space $\mathbf{H} = \sum_{r\in\Omega} \oplus L_r^2$. Also let the corresponding boundary complex* *symplectic space be* $\mathsf{S} = \mathbf{D}(\mathbf{T}_1)/\mathbf{D}(\mathbf{T}_0)$, *with the symplectic product* $[\cdot : \cdot]_{\mathsf{S}}$ *and Hilbert space scalar product* $\langle \cdot, \cdot \rangle_{\mathsf{S}}$, *as above.*

Then there exists a natural bijective correspondence between the set $\{\mathbf{T}\}$ *of all self-adjoint operators* $\mathbf{T}$ *on domains* $\mathbf{D}(\mathbf{T}) \subset \mathbf{H}$, *generated by the multi-interval system as extensions of* $\mathbf{T}_0$ *(hence restrictions of* $\mathbf{T}_1$ *on* $\mathbf{D}(\mathbf{T}_1)$*), and the set* $\{\mathsf{L}\}$ *of all complete Lagrangian subspaces* L *of the boundary space* S.

Namely, each such self-adjoint operator $\mathbf{T}$ *on* $\mathbf{D}(\mathbf{T})$ *corresponds to the complete Lagrangian*

$$\mathsf{L} := \mathbf{D}(\mathbf{T})/\mathbf{D}(\mathbf{T}_0) \text{ or } \mathsf{L} := \Psi\mathbf{D}(\mathbf{T})$$

in terms of the natural projection map $\Psi : \mathbf{D}(\mathbf{T}_1) \to \mathsf{S}$, *as defined in (5.22). Conversely, for each complete Lagrangian space* $\mathsf{L} \subset \mathsf{S}$ *the corresponding self-adjoint extension* $\mathbf{T}$ *of* $\mathbf{T}_0$ *is defined by*

$$\mathbf{T}\mathbf{f} := \mathbf{T}_1\mathbf{f} \text{ for all } \mathbf{f} \in \mathbf{D}(\mathbf{T})$$

where the domain of $\mathbf{T}$ *is*

$$\mathbf{D}(\mathbf{T}) := \{\mathbf{f} \in \mathbf{D}(\mathbf{T}_1) : [\mathsf{f} : \mathsf{L}] = 0\}$$

or, equivalently

$$\mathbf{D}(\mathbf{T}) := \{\mathbf{f} \in \mathbf{D}(\mathbf{T}_1) : \mathsf{f} \in \mathsf{L}\}$$

with, as before, $\mathsf{f} = \Psi\mathbf{f} = \{\mathbf{f} + \mathbf{D}(\mathbf{T}_0)\}$.

Indeed for any choice of orthonormal basis $\{\mathsf{b}_k : k \in \Omega_0\}$ *of the Hilbert subspace* L *of* S, *there is the expansion*

$$\mathbf{D}(\mathbf{T}) = \sum_{k\in\Omega_0} c_k\mathbf{b}_k + \mathbf{D}(\mathbf{T}_0)$$

where $\mathbf{b}_k \in \mathbf{N}^- \oplus \mathbf{N}^+$ *with* $\Psi\mathbf{b}_k = \mathsf{b}_k$ *for all* $k \in \Omega_0$ *(an index set with card*$(\Omega_0) = \dim(\mathsf{L})$, *see Remark 6.4) and numbers* $c_k \in \mathbb{C}$ *with* $\sum_{k\in\Omega_0} |c_k|^2 < +\infty$. *Note that* $\{\mathbf{b}_k : k \in \Omega_0\}$ *is uniquely determined from the chosen set* $\{\mathsf{b}_k : k \in \Omega_0\}$, *and is an orthonormal set in* $\mathbf{N}^- \oplus \mathbf{N}^+$.

In more detail, for each set of numbers $\{c_k \in \mathbb{C} : k \in \Omega_0\}$ *with* $\sum_{k\in\Omega_0} |c_k|^2 < +\infty$ *and each* $\mathbf{g}_0 \in \mathbf{D}(\mathbf{T}_0)$ *then* $\mathbf{g} \in \mathbf{D}(\mathbf{T})$ *when defined by*

$$\mathbf{g} := \sum_{k\in\Omega_0} c_k\mathbf{b}_k + \mathbf{g}_0$$

with convergence in the $\mathbf{D}$*-norm. Moreover, each* $\mathbf{h} \in \mathbf{D}(\mathbf{T})$ *has a unique expansion of this form.*

PROOF. See [**10**] where a generalization of this Theorem is proved within an abstract Hilbert space context. The uniqueness of the expansion for $\mathbf{g} \in \mathbf{D}(\mathbf{T})$ follows because $\mathbf{g} = \mathbf{g}_0 + \mathbf{g}_- + \mathbf{g}_+$, and $\Psi(\mathbf{g}_- + \mathbf{g}_+) \in \mathsf{L}$ has a unique such expansion in terms of the given orthonormal basis $\{\mathsf{b}_k : k \in \Omega_0\}$ for L. Finally $\mathbf{g} - \sum_{k\in\Omega_0} c_k\mathbf{b}_k = \mathbf{g}_0$ as required. □

The next Corollary also follows from the classical results of Stone-von Neumann [**4**, Chapter XII.4.13].

COROLLARY 6.1. *Under the conditions of Theorem 6.2 the set* $\{\mathbf{T}\}$ *is nonempty if and only if there exists a complete Lagrangian subspace* $\mathsf{L} \subset \mathsf{S}$. *This situation occurs if and only if the deficiency indices* $\{\mathbf{d}^-, \mathbf{d}^+\}$ *of the minimal operator* $\mathbf{T}_0$ *are equal cardinal numbers,* $\mathbf{d}^- = \mathbf{d}^+$, *in which case we write* $\mathbf{d} = \mathbf{d}^\pm$.

Remark 6.3. In the case when $\mathbf{H}$ is a separable Hilbert space, that is when the index set Ω is finite or denumerably infinite, say $\operatorname{card}(\Omega) = N$ or $\aleph_0$, then

$$\mathbf{d}^{\pm} = \sum_{r=1}^{N} d_r^{\pm} \text{ or } \sum_{r=1}^{\infty} d_r^{\pm}. \tag{6.31}$$

Thus the requirement for the existence of self-adjoint extensions $\mathbf{T}$ is that $\mathbf{d}^- = \mathbf{d}^+$, including the case when both the infinite series in (6.31) diverge so that $\mathbf{d}^- = \mathbf{d}^+ = \aleph_0$. This special case was first considered in [**15**].

In the case when Ω is finite, see the early results in [**14**], say $\Omega = \{1, 2, \ldots, N\}$, then $\dim(\mathsf{S}) < +\infty$ and this necessitates that (see (2.23))

$$\dim(\mathsf{S}) = 2\mathbf{d} \text{ and } \dim(\mathsf{L}) = \mathbf{d}, \text{ with } \mathbf{d} \leq \sum_{r=1}^{N} n_r.$$

However when S is infinite dimensional, then

$$\dim(\mathsf{S}) = \dim(\mathsf{L}) = \mathbf{d}, \text{ with } \mathbf{d} \geq \aleph_0,$$

as is clarified below.

We now return to the analysis of the general multi-interval system $\{I_r, M_r, w_r : r \in \Omega\}$, as in (6.1) and Definition 2.1 above, for further study of the deficiency indices $\{\mathbf{d}^-, \mathbf{d}^+\}$. The boundary complex space $\mathsf{S} = \mathbf{D}(\mathbf{T}_1)/\mathbf{D}(\mathbf{T}_0)$ with the symplectic product $[\mathsf{f} : \mathsf{g}]_{\mathsf{S}} = [\mathbf{f} : \mathbf{g}]_{\mathbf{D}}$ (where $\mathsf{f} = \{\mathbf{f} + \mathbf{D}(\mathbf{T}_0)\}$, $\mathsf{g} = \{\mathbf{g} + \mathbf{D}(\mathbf{T}_0)\}$ with $\mathbf{f}, \mathbf{g} \in \mathbf{N}^- \oplus \mathbf{N}^+$), is also a Hilbert space with norm given by

$$\|\mathsf{f}\|_{\mathsf{S}}^2 = \|\mathbf{f}_-\|_{\mathbf{D}}^2 + \|\mathbf{f}_+\|_{\mathbf{D}}^2 = 2\|\mathbf{f}_-\|_{\mathbf{H}}^2 + 2\|\mathbf{f}_+\|_{\mathbf{H}}^2 \text{ with } \mathbf{f} = \mathbf{f}_- + \mathbf{f}_+ \text{ where } \mathbf{f}_{\pm} \in \mathbf{N}^{\pm}.$$

Further, for each fixed $r \in \Omega$, we can define the analogous spaces $\mathsf{S}_r = D(T_{1,r})/D(T_{0,r})$ of cosets $\mathsf{f}_r = \{f_r + D(T_{0,r})\}$, $\mathsf{g}_r = \{g_r + D(T_{0,r})\}$ with $f_r, g_r \in N_r^- \oplus N_r^+$, and then compute the corresponding symplectic products and norms in S_r by the formulae (see Remark 6.1)

$$[\mathsf{f}_r : \mathsf{g}_r]_{\mathsf{S}_r} = [f_r : g_r]_r = \langle T_{1,r} f_r, g_r \rangle_r - \langle f_r, T_{1,r} g_r \rangle_r$$

$$\|\mathsf{f}_r\|_{\mathsf{S}_r}^2 = \|f_{r-} + f_{r+}\|_{D_r}^2 = 2\|f_{r-}\|_r^2 + 2\|f_{r+}\|_r^2.$$

We note that $N_r^{\pm}$ and S_r are each a Hilbert space (finite dimensional), and hence we can construct the corresponding direct sum Hilbert spaces, over $r \in \Omega$, just as for the system Hilbert space $\mathbf{H} = \sum_{r \in \Omega} \oplus L_r^2$, as in Definition 5.1. We shall show that such infinite direct sums can be used to construct $\sum_{r\in} \oplus \mathsf{S}_r$, and that the symplectic structure is treated as well as the metric norm structure (compare Definition 3.2 for finite direct sum decompositions).

Definition 6.2. For each $r \in \Omega$ the complex Hilbert space N_r^- has the norm indicated by, using a notation for the space D_r that is consistent with the definitions (6.11) and (6.12),

$$\|f_{r-}\|_{D_r}^2 := 2\|f_{r-}\|_r^2 \text{ and the symplectic product } [f_{r-} : g_{r-}]_r = -2i \langle f_{r-}, g_{r-} \rangle_r.$$

Accordingly we define the direct sum Hilbert space

$$\mathbf{N}_{\oplus}^- := \sum_{r \in \Omega} \oplus N_r^- \text{ with } \|\mathbf{f}_{\oplus}\|_{\oplus}^2 = 2\sum_{r \in \Omega} \|f_{r-}\|_r^2$$

where

$$\mathbf{f}_{\oplus} := \left\{ f_{r-} \in N_r^- : \sum_{r\in\Omega} \|f_{r-}\|_r^2 < +\infty \right\} \in \mathbf{N}_{\oplus}^-.$$

Also the symplectic product of $\mathbf{f}_{\oplus}, \mathbf{g}_{\oplus} \in \mathbf{N}_{\oplus}^-$ is

$$[\mathbf{f}_{\oplus} : \mathbf{g}_{\oplus}]_{\oplus} := \sum_{r\in\Omega} [f_{r-} : g_{r-}]_r = -2i \sum_{r\in\Omega} \langle f_{r-}, g_{r-} \rangle_r$$

(absolutely convergent sum, with only countably many non-zero terms).

In a similar way we define the complex Hilbert space

$$\mathbf{N}_{\oplus}^+ := \sum_{r\in\Omega} \oplus N_r^+ \text{ with } \|\mathbf{f}_{\oplus}\|_{\oplus}^2 = 2 \sum_{r\in\Omega} \|f_{r+}\|_r^2$$

and the corresponding symplectic product indicated by

$$[\mathbf{f}_{\oplus} : \mathbf{g}_{\oplus}]_{\oplus} := \sum_{r\in\Omega} [f_{r+} : g_{r+}]_r = 2i \sum_{r\in\Omega} \langle f_{r+}, g_{r+} \rangle_r \,.$$

Then both $\mathbf{N}_{\oplus}^{\pm}$ are Hilbert spaces and complex symplectic spaces.

LEMMA 6.2. *We have*

$$\mathbf{N}^- = \sum_{r\in\Omega} \oplus N_r^- \text{ and } \mathbf{N}^+ = \sum_{r\in\Omega} \oplus N_r^+$$

both as Hilbert spaces and as complex symplectic spaces.

PROOF. The Hilbert space $\mathbf{N}^+$ consists of vectors $\mathbf{f} \in \mathbf{D}(\mathbf{T}_1)$ with $\mathbf{T}_1\mathbf{f} = i\mathbf{f}$. Thus $\mathbf{f} = \{f_r \in D(T_{1,r}) : r \in \Omega\}$ with $\sum_{r\in\Omega} \|f_r\|_r^2 < +\infty$ and $T_{1,r}f_r = if_r$ for all $r \in \Omega$, and so $\sum_{r\in\Omega} \|T_{1,r}f_r\|_r^2 < +\infty$. Hence each component $f_r \in N_r^+$, and we conclude that $\mathbf{N}^+ \subseteq \mathbf{N}_{\oplus}^+ = \sum_{r\in\Omega} \oplus N_r^+$. Conversely each vector of $\mathbf{N}_{\oplus}^+$ has the form $\{f_r \in N_r^+ : r \in \Omega\}$ with $\sum_{r\in\Omega} \|f_r\|_r^2 < +\infty$, and so belongs to $\mathbf{N}^+$; thus $\mathbf{N}_{\oplus}^+ \subseteq \mathbf{N}^+$. Hence $\mathbf{N}^+ = \mathbf{N}_{\oplus}^+$ as Hilbert spaces; similarly $\mathbf{N}^- = \mathbf{N}_{\oplus}^-$.

It is now straightforward to compute the equality of the corresponding symplectic products, and we omit the direct calculations. □

THEOREM 6.3. *Let $\{I_r, M_r, w_r : r \in \Omega\}$ be a multi-interval system, as in Definition 2.1, and consider the boundary complex symplectic space $\mathsf{S} = \mathbf{D}(\mathbf{T}_1)/\mathbf{D}(\mathbf{T}_0)$ with symplectic product $[\cdot : \cdot]_{\mathsf{S}}$; also with Hilbert space scalar product $\langle \cdot, \cdot \rangle_{\mathsf{S}}$ and norm $\|\cdot\|_{\mathsf{S}}$ as before. For each fixed $r \in \Omega$ let $\mathsf{S}_r = D(T_{1,r})/D(T_{0,r})$ with analogous symplectic and scalar products $[\cdot : \cdot]_r$ and $\langle \cdot, \cdot \rangle_r$, as above.*

Define the Hilbert space direct sum

$$\mathsf{S}_{\oplus} := \sum_{r\in\Omega} \oplus \mathsf{S}_r \text{ with vectors } \mathsf{h}_{\oplus} = \{\mathsf{h}_r \in \mathsf{S}_r : r \in \Omega\}$$

with norm given by

$$\|\mathsf{h}_{\oplus}\|_{\mathsf{S}_{\oplus}}^2 := \sum_{r\in\Omega} \|\mathsf{h}_r\|_{\mathsf{S}_r}^2 < +\infty.$$

Then $\mathsf{S}_{\oplus}$ is a Hilbert space, and also a complex symplectic space with

$$[\mathsf{g}_{\oplus} : \mathsf{h}_{\oplus}]_{\mathsf{S}_{\oplus}} = \sum_{r\in\Omega} [\mathsf{g}_r : \mathsf{h}_r]_{\mathsf{S}_r} \text{ for } \mathsf{g}_{\oplus}, \mathsf{h}_{\oplus} \in \mathsf{S}_{\oplus}$$

where the infinite series is absolutely convergent with only a countable number of non-zero terms.

Define the linear map

$$\Phi : \mathsf{S} \to \mathsf{S}_{\oplus} = \sum_{r\in\Omega} \oplus \mathsf{S}_r \textit{ by } \mathsf{f} \to \mathsf{f}_{\oplus}$$

for

$$\mathsf{f} = \{\mathbf{f}_- + \mathbf{f}_+ + \mathbf{D}(\mathbf{T}_0)\} \to \mathsf{f}_{\oplus} = \{f_{r-} + f_{r+} + D(T_{0,r}) : r \in \Omega\}$$

where $\mathbf{f}_{\pm} = \{f_{r\pm} \in N_r^{\pm} : r \in \Omega\} \in \mathbf{N}^{\pm}$. *Then* Φ *is a Hilbert space isomorphism (unitary bijection) of* S *onto* $\mathsf{S}_{\oplus}$, *and* Φ *is also a symplectic isomorphism. Accordingly, we identify* S *and* $\mathsf{S}_{\oplus}$, *under* Φ, *and write*

$$\mathsf{S} = \sum_{r\in\Omega} \oplus \mathsf{S}_r.$$

PROOF. For a given $\mathsf{f} = \{\mathbf{f}_- + \mathbf{f}_+ + \mathbf{D}(\mathbf{T}_0)\}$, with $\mathbf{f}_{\pm} \in \mathbf{N}^{\pm}$ so that $\mathsf{f} \in \mathsf{S}$, define $\Phi\mathsf{f} := \mathsf{f}_{\oplus} = \{f_{r-} + f_{r+} + D(T_{0,r}) : r \in \Omega\}$; we shall show that $\mathsf{f}_{\oplus} \in \mathsf{S}_{\oplus}$. Now the components of $\mathbf{f}_{\pm}$ are precisely given by $\mathbf{f}_{\pm} = \{f_{r-} + f_{r+} : r \in \Omega\}$, see Lemma 6.2. Then the corresponding norms are given by

$$\|\mathsf{f}\|_{\mathsf{S}}^2 = 2\,\|\mathbf{f}_-\|_{\mathbf{H}}^2 + 2\,\|\mathbf{f}_+\|_{\mathbf{H}}^2$$

and

$$\|\mathsf{f}_{\oplus}\|_{\mathsf{S}_{\oplus}}^2 = \sum_{r\in\Omega} \left(\|f_{r-}\|_{D_r}^2 + \|f_{r+}\|_{D_r}^2\right) = 2\,\|\mathbf{f}_-\|_{\mathbf{H}}^2 + 2\,\|\mathbf{f}_+\|_{\mathbf{H}}^2\,.$$

Hence Φ is an injective unitary map into $\mathsf{S}_{\oplus}$.

However, given any $\mathsf{g}_{\oplus} = \{g_{r-} + g_{r+} + D(T_{0,r}) : r \in \Omega\} \in \mathsf{S}_{\oplus}$, then $g_{r\pm} \in N_r^{\pm}$ and, also, $\sum_{r\in\Omega} \|g_{r\pm}\|_{D_r}^2 = 2\,\|\mathbf{g}_{\pm}\|_{\mathbf{H}}^2 < +\infty$. From this construction we now define

$$\mathsf{g} := \{\mathbf{g}_- + \mathbf{g}_+ + \mathbf{D}(\mathbf{T}_0)\} \in \mathsf{S},$$

and note that $\Phi\mathsf{g} = \mathsf{g}_{\oplus}$, so that Φ is a unitary surjection onto $\mathsf{S}_{\oplus}$.

Next we examine the symplectic products in S and $\mathsf{S}_{\oplus}$. Take vectors $\mathsf{g} = \{\mathbf{g}_- + \mathbf{g}_+ + \mathbf{D}(\mathbf{T}_0)\}$ and $\mathsf{h} = \{\mathbf{h}_- + \mathbf{h}_+ + \mathbf{D}(\mathbf{T}_0)\}$ in S; let $\mathsf{g}_{\oplus} = \Phi\mathsf{g}$ and $\mathsf{h}_{\oplus} = \Phi\mathsf{h}$, hence both in $\mathsf{S}_{\oplus}$. Then, using the same notation as above,

$$\begin{aligned}[\mathsf{g} : \mathsf{h}]_{\mathsf{S}_{\oplus}} &= \sum_{r\in\Omega} [g_{r-} + g_{r+} : h_{r-} + h_{r+}]_r = \sum_{r\in\Omega} [g_{r-} : h_{r-}]_r + \sum_{r\in\Omega} [g_{r+} : h_{r+}]_r \\ &= [\mathbf{g}_- : \mathbf{h}_-]_{\mathbf{D}} + [\mathbf{g}_+ : \mathbf{h}_+]_{\mathbf{D}} = [\mathsf{g} : \mathsf{h}]_{\mathsf{S}}\end{aligned}$$

Therefore Φ is a symplectic isomorphism, as asserted. □

COROLLARY 6.2. *Let* $\mathbf{N}^- = \sum_{r\in\Omega} \oplus N_r^-$, *and select an orthonormal basis for each* N_r^-. *Then the union of all the vectors of all these bases constitutes an orthonormal basis for* $\mathbf{N}^-$. *Therefore*

$$\dim(\mathbf{N}^-) = \mathrm{card}\left\{\bigcup_{r\in\Omega} \{\textit{selected orthonormal basis for } N_r^-\}\right\}$$

and so $\mathbf{d}^- = \dim(\mathbf{N}^-) = \sum_{r\in\Omega} d_r^-$, *as the cardinal number sum.*

A similar result holds for $\mathbf{d}^+ = \dim(\mathbf{N}^+) = \sum_{r\in\Omega} d_r^+$.

PROOF. If a vector $\mathbf{f} = \{f_r : r \in \Omega\} \in \mathbf{N}^-$ is orthogonal to every Hilbert subspace N_r^-, then $\mathbf{f} = 0$. □

COROLLARY 6.3. *We have*

$$\dim(\mathsf{S}) = \dim(\mathbf{N}^-) + \dim(\mathbf{N}^+) = \mathbf{d}^- + \mathbf{d}^+,$$

as the sum of cardinal numbers (Hilbert space dimensions).

If $\dim(\mathsf{S}_r)$ *is a positive integer for each* $r \in \Omega$, *then* $\dim(\mathsf{S}) < +\infty$ *if and only if* $\mathrm{card}(\Omega) < +\infty$, *and* $\dim(\mathsf{S}) = \mathrm{card}(\Omega)$ *in case* $\mathrm{card}(\Omega)$ *is infinite.*

PROOF. Clear. □

In Section 3 above we define the symplectic invariants D, p, q, Δ and Ex for a finite dimensional symplectic space S of dimension $D \geq 1$; see Definition 3.4 and (3.16). In Section 4 we apply these concepts to the single-interval system, see (4.9) to (4.11); in particular these invariants are related to the deficiency indices $\{d^-, d^+\}$ of the system. We now present an introductory sketch of some definitions and properties for intrinsic invariants of infinite dimensional abstract symplectic spaces, as illustrated by $\mathsf{S} \approx \mathbf{N}^- \oplus \mathbf{N}^+$ as in Theorem 6.1 above; for full details see the results in [**11**].

Definition 6.3. Let S be a complex symplectic space; then a linear subspace $\mathsf{V} \subseteq \mathsf{S}$ is a *negativity space* in case:

$$\mathrm{Im}\left([\mathsf{u} : \mathsf{u}]_{\mathsf{S}}\right) < 0 \text{ for all } \mathsf{u} \in \mathsf{V} \text{ with } \mathsf{u} \neq 0,$$

and the norm defined by $\|\mathsf{u}\|_{\mathsf{V}}^2 := i[\mathsf{u} : \mathsf{u}]_{\mathsf{S}}$, for all $\mathsf{u} \in \mathsf{V}$, defines V as a (complete) Hilbert space.

Similarly $\mathsf{V} \subseteq \mathsf{S}$ is a *positivity space* in case

$$\mathrm{Im}\left([\mathsf{u} : \mathsf{u}]_{\mathsf{S}}\right) > 0 \text{ for all } \mathsf{u} \in \mathsf{V} \text{ with } \mathsf{u} \neq 0$$

and also $\|\mathsf{u}\|_{\mathsf{V}}^2 := -i[\mathsf{u} : \mathsf{u}]_{\mathsf{S}}$ defines V as a (complete) Hilbert space.

Note. Let the space $\mathsf{S} = \mathbf{D}(\mathbf{T}_1)/\mathbf{D}(\mathbf{T}_0)$ be defined as in Definitions 5.4 and 6.1. From the prior results the deficiency subspaces $\mathbf{N}^-$ and $\mathbf{N}^+$ are examples of negativity and positivity subspaces of this complex symplectic space (Hilbert space) S. However for this space S there can be other negativity and positivity spaces V which are not contained in $\mathbf{N}^{\pm}$. In this definition we refer to certain closed subspaces of the Hilbert space S, but the results are unchanged if we consider complex symplectic subspaces closed in the symplectic weak topology of S, see Remark 6.2 and [**11**].

Definition 6.4. Let $\mathsf{S} \approx \mathbf{N}^- \oplus \mathbf{N}^+$ be a complex symplectic space, as before. Then define the symplectic invariants for S by:

$$\mathbf{p} := \sup\{\dim(\mathbf{P}) : \ \mathbf{P} \text{ is a positivity subspace of } \mathsf{S}\}$$

$$\mathbf{q} := \sup\{\dim(\mathbf{N}) : \ \mathbf{N} \text{ is a negativity subspace of } \mathsf{S}\}$$

$$\boldsymbol{\Delta} := \sup\{\dim(\mathsf{L}) : \mathsf{L} \text{ is a Lagrangian subspace of } \mathsf{S}\}.$$

The positivity index $\mathbf{p}$, negativity index $\mathbf{q}$, and the Lagrangian index $\boldsymbol{\Delta}$ are cardinal numbers whose existence is guaranteed by the well-ordering property for sets of cardinals.

It can be demonstrated that [**11**]

$$\mathbf{p} = \mathbf{d}^+ = \dim(\mathbf{N}^+) \text{ and } \mathbf{q} = \mathbf{d}^- = \dim(\mathbf{N}^-)$$

and

$$\dim(\mathsf{S}) = \mathbf{p} + \mathbf{q} \text{ and } \mathbf{\Delta} = \min\{\mathbf{p}, \mathbf{q}\}.$$

Also for infinite dimensional symplectic spaces S

$$\dim(\mathsf{S}) = \max\{\mathbf{p}, \mathbf{q}\}.$$

In the infinite dimensional case the excess invariant $Ex \equiv Ex(\mathsf{S})$ cannot be defined numerically, but in terms of cardinal numbers we can introduce the notation:

$$Ex(\mathsf{S}) > 0 \text{ signifies } \mathbf{p} > \mathbf{q}$$
$$Ex(\mathsf{S}) < 0 \text{ signifies } \mathbf{p} < \mathbf{q}$$
$$Ex(\mathsf{S}) = 0 \text{ signifies } \mathbf{p} = \mathbf{q}.$$

With these considerations in mind we assert the theorem:

THEOREM 6.4. *Let* $\mathsf{S} = \mathbf{D}(\mathbf{T}_1)/\mathbf{D}(\mathbf{T}_0) \approx \mathbf{N}^- \oplus \mathbf{N}^-$, *as before. Then there exists a complete Lagrangian subspace* $\mathsf{L} \subset \mathsf{S}$ *if and only if the symplectic invariants of* S *satisfy*

$$\mathbf{p} = \mathbf{q} \textit{ or, equivalently, } Ex(\mathsf{S}) = 0.$$

In this case

$$\dim(\mathsf{L}) = \mathbf{\Delta} = \mathbf{p} = \mathbf{q} = \begin{cases} \dim(\mathsf{S}) \textit{ when infinite} \\ \frac{1}{2}\dim(\mathsf{S}) \textit{ when finite.} \end{cases}$$

PROOF. For details see [**11**]. □

Finally we assert a result relating the symplectic invariants $\mathbf{p}, \mathbf{q}, \mathbf{\Delta}$ of S, as in Definition 6.4, to the corresponding invariants p_r, q_r, Δ_r of S_r, for each $r \in \Omega$, as in Theorem 6.3.

PROPOSITION 6.1. *Let* $\{I_r, M_r, w_r : r \in \Omega\}$ *be a multi-interval system, as in Definition* 2.1, *and consider the boundary complex symplectic space*

$$\mathsf{S} = \mathbf{D}(\mathbf{T}_1)/\mathbf{D}(\mathbf{T}_0),$$

with symplectic product $[\cdot : \cdot]_{\mathsf{S}}$ *and also with the Hilbert space scalar product* $\langle \cdot, \cdot \rangle_{\mathsf{S}}$ *and norm* $\|\cdot\|_{\mathsf{S}}$. *As in Theorem* 6.3, $\mathsf{S} = \sum_{r\in\Omega} \oplus \mathsf{S}_r$.

Then the corresponding symplectic invariants are related by

$$\mathbf{p} = \sum_{r\in\Omega} p_r \qquad \mathbf{q} = \sum_{r\in\Omega} q_r \qquad \dim(\mathsf{S}) = \sum_{r\in\Omega} \dim(\mathsf{S}_r);$$

also

$$\mathbf{\Delta} \geq \sum_{r\in\Omega} \Delta_r$$

and $Ex(\mathsf{S}_r) = 0$ *for all* $r \in \Omega$ *implies equality. In these results all sums, equalities and inequalities are in the sense of cardinal numbers.*

PROOF. From Corollary 6.2 above $\mathbf{d} = \sum_{r\in\Omega} d_r^{\pm}$. Then the equalities $\mathbf{p} = \mathbf{d}^+, \mathbf{q} = \mathbf{d}^-, p_r = d_r^+, q_r = d_r^-$ imply the conclusions for $\mathbf{p}, \mathbf{q}, \dim(\mathsf{S})$. The additional details for the proof are found in [**11**]. □

Example 6.1. In contrast to the results in Theorem 3.1 on finite dimensional symplectic spaces, we here consider infinite dimensional complex symplectic spaces $\mathsf{S} = \sum_{r=1}^{\infty} \oplus \mathsf{S}_r$, as in Proposition 6.1 with $\Omega = \mathbb{N}$. If we take

$$p_r = \begin{cases} 1 \text{ for } r \text{ even} \\ 0 \text{ for } r \text{ odd} \end{cases} \qquad q_r = \begin{cases} 0 \text{ for } r \text{ even} \\ 1 \text{ for } r \text{ odd} \end{cases},$$

then all $\Delta_r = 0$, yet $\mathbf{p} = \mathbf{q} = \mathbf{\Delta} = \aleph_0$ so $\dim(\mathsf{S}) = \aleph_0$ and $Ex(\mathsf{S}) = 0$.

On the other hand if we take $p_r = 2$, $q_r = 1$ for all $r \in \mathbb{N}$, then

$$\mathbf{p} = \mathbf{q} = \mathbf{\Delta} = \sum_{r=1}^{\infty} \Delta_r = \aleph_0$$

so $Ex(\mathsf{S}) = 0$ even though $Ex(\mathsf{S}_r) > 0$ for all $r \in \mathbb{N}$.

We next review a previous version of the GKN-Theorem, proved for countable index sets Ω in [**15**] (see also [**13**]), and relate the prior concept of *maximal GKN-sets* to the constructions above dealing with complete Lagrangians of the boundary complex symplectic space S of $\{I_r, M_r, w_r : r \in \Omega\}$. Again the following results allow Ω to have any cardinality, finite, denumerable, or greater than $\aleph_0$.

Remark 6.4. We introduce an index set Ω_0, which is specific to each appearance and may vary in different cases; moreover Ω_0 is not related to Ω in any special way.

Definition 6.5. Let $\{I_r, M_r, w_r : r \in \Omega\}$ be a multi-interval system, with maximal and minimal operators $\mathbf{T}_1$ on $\mathbf{D}(\mathbf{T}_1)$ and $\mathbf{T}_0$ on $\mathbf{D}(\mathbf{T}_0)$, respectively, in the Hilbert space $\mathbf{H} = \sum_{r\in\Omega} \oplus L_r^2$, as before. Consider the boundary complex symplectic space, as in Theorem 6.1,

$$\mathsf{S} = \mathbf{D}(\mathbf{T}_1)/\mathbf{D}(\mathbf{T}_0) \approx \mathbf{N}^- \oplus \mathbf{N}^+ \tag{6.32}$$

with the corresponding symplectic product $[\cdot : \cdot]_{\mathsf{S}}$, written here as $[\cdot : \cdot]$ for simplicity, and the Hilbert space norm $\|\cdot\|_{\mathsf{S}}$, as in Definition 6.1.

A set $\{\mathsf{b}_k : k \in \Omega_0\}$ is defined to be a *GKN-set* in S in case

(6.33)

(*i*) $\{\mathsf{b}_k : k \in \Omega_0\}$ is a locally linearly independent subset of S, *i.e.* every finite (non-trivial) linear combination over $\mathbb{C}$ is non-zero

(*ii*) $\mathsf{L}_0 := \operatorname{span}\{\mathsf{b}_k : k \in \Omega_0\}$ is a Lagrangian submanifold of S (here L_0 consists of all finite linear combinations over $\mathbb{C}$).

Remark 6.5 An equivalent formulation of (*ii*) above is the assumption that $[\mathsf{b}_{k_1} : \mathsf{b}_{k_2}] = 0$ for all pairs taken from the set $\{\mathsf{b}_k : k \in \Omega_0\}$, see [**15**, Section 7].

In addition $\{\mathsf{b}_k : k \in \Omega_0\}$ is defined to be a *maximal GKN-set* of S in case (i) and (ii) are augmented by the *maximal condition*

(*iii*) If $\mathsf{b} \in \mathsf{S}$ with $[\mathsf{b} : \mathsf{L}_0] = 0$ and $\mathsf{b} \in \mathsf{L}_0^{\perp}$, then $\mathsf{b} = 0$.

THEOREM 6.5. *Let $\{I_r, M_r, w_r : r \in \Omega\}$ be a multi-interval system, with boundary space*

$$\mathsf{S} = \mathbf{D}(\mathbf{T}_1)/\mathbf{D}(\mathbf{T}_0) \approx \mathbf{N}^- \oplus \mathbf{N}^+.$$

Then S is a complex symplectic space with

$$[\mathsf{u} : \mathsf{v}] = [\mathbf{u}_- : \mathbf{v}_-]_{\mathbf{D}} + [\mathbf{u}_+ : \mathbf{v}_+]_{\mathbf{D}} \tag{6.34}$$

and a Hilbert space with

$$\langle \mathsf{u},\mathsf{v}\rangle_{\mathsf{S}} = \langle \mathbf{u}_-,\mathbf{v}_-\rangle_{\mathbf{D}} + \langle \mathbf{u}_+,\mathbf{v}_+\rangle_{\mathbf{D}} \tag{6.35}$$

for all $\mathsf{u} \approx \mathbf{u}_- + \mathbf{u}_+, \mathsf{v} \approx \mathbf{v}_- + \mathbf{v}_+ \in \mathsf{S}$, *as in Theorem* 6.1 *and Definition* 6.1 *above.*

For each $\{\mathsf{b}_k : k \in \Omega_0\}$, *a maximal GKN-set in* S, *let* $\mathsf{L}_0 = \operatorname{span}\{\mathsf{b}_k : k \in \Omega_0\}$ *and define* L *as the closure of* L_0 *in this Hilbert space, i.e.*

$$\mathsf{L} := \overline{\mathsf{L}_0} = \overline{\operatorname{span}}\{\mathsf{b}_k : k \in \Omega_0\}. \tag{6.36}$$

Then L *is a complete Lagrangian subspace of* S, *and moreover*

$$\mathsf{L} = \{\mathsf{f} \in \mathsf{S} : [\mathsf{f} : \mathsf{b}_k] = 0 \text{ for all } k \in \Omega_0\}. \tag{6.37}$$

On the other hand, let L *be a complete Lagrangian subspace of* S *(hence a closed Hilbert space of* S*). Then there exists a maximal GKN-set* $\{\mathsf{h}_k : k \in \Omega_0\}$ *in* S *with*

$$\overline{\operatorname{span}}\{\mathsf{h}_k : k \in \Omega_0\} = \mathsf{L}.$$

In fact, each orthonormal basis for the Hilbert space L *is such a maximal GKN-set.*

PROOF. Let $\{\mathsf{b}_k\}$ be a maximal GKN-set in S, hence satisfying the conditions (i), (ii) and (iii) of Definition 6.5. Since $\mathsf{L}_0 := \operatorname{span}\{\mathsf{b}_k : k \in \Omega_0\}$ is a Lagrangian submanifold in the Hilbert space S, we conclude that its closure $\overline{\mathsf{L}_0}$ is also a Lagrangian subspace in S, because of the continuity of the given symplectic product $[\cdot : \cdot]$. Defining $\mathsf{L} := \overline{\mathsf{L}_0}$ we must prove that L is a complete Lagrangian subspace.

Now the Hilbert space S has an orthogonal direct sum decomposition $\mathsf{S} = \mathsf{L} \oplus \mathsf{L}^\perp$ and each vector $\mathsf{u} \in \mathsf{S}$ has a corresponding unique decomposition $\mathsf{u} = \mathsf{u}_{\mathsf{L}} + \mathsf{u}_{\mathsf{L}^\perp}$.

Assume that $[\mathsf{u} : \mathsf{L}] = 0$. Then $[\mathsf{u}_{\mathsf{L}} + \mathsf{u}_{\mathsf{L}^\perp} : \mathsf{L}] = 0$ and so $[\mathsf{u}_{\mathsf{L}^\perp} : \mathsf{L}] = 0$. Thus $[\mathsf{u}_{\mathsf{L}^\perp} : \mathsf{L}_0] = 0$ and $\mathsf{u}_{\mathsf{L}^\perp} \in \mathsf{L}_0^\perp$. From property *(iii)* of the GKN-set $\{\mathsf{b}_k : k \in \Omega_0\}$, we conclude that $\mathsf{u}_{\mathsf{L}^\perp} = 0$ and therefore $\mathsf{u} = \mathsf{u}_{\mathsf{L}} \in \mathsf{L}$, which proves that L is a complete Lagrangian subspace.

Now consider the reverse argument, and assume that $\mathsf{L} \subset \mathsf{S}$ is a given complete Lagrangian subspace. Then L is a closed Hilbert subspace of S, and let $\{\mathsf{h}_k : k \in \Omega_0\}$ be any orthonormal basis of L. Then $\{\mathsf{h}_k : k \in \Omega_0\}$ certainly satisfies conditions *(i)* and *(ii)* of Definition 6.5.

In order to verify condition *(iii)* take $\mathsf{b} \in \mathsf{S}$, with $\mathsf{L}_0 := \operatorname{span}\{\mathsf{h}_k : k \in \Omega_0\}$, such that

$$[\mathsf{b} : \mathsf{h}_k] = 0 \text{ for all } k \in \Omega_0, \textit{ i.e. } [\mathsf{b} : \mathsf{L}_0] = 0.$$

By continuity $[\mathsf{b} : \mathsf{L}] = 0$, since $\mathsf{L} = \overline{\operatorname{span}}\{\mathsf{h}_k : k \in \Omega_0\} = \overline{\mathsf{L}_0}$. Because we are now assuming that L is a complete Lagrangian of S, $\mathsf{b} \in \mathsf{L}$. But if we also assume that $\mathsf{b} \in \mathsf{L}^\perp$ (or merely $\mathsf{b} \in \mathsf{L}_0^\perp$), then $\mathsf{b} = 0$, and thus the maximality condition *(iii)* is satisfied. Therefore $\{\mathsf{h}_k : k \in \Omega_0\}$ is a maximal GKN-set in S, as required, and $\overline{\operatorname{span}}\{\mathsf{h}_k : k \in \Omega_0\} = \mathsf{L}$. □

COROLLARY 6.4. *Let* $\{I_r, M_r, w_r : r \in \Omega\}$ *be given with*

$$\mathsf{S} = \mathbf{D}(\mathbf{T}_1)/\mathbf{D}(\mathbf{T}_0) \approx \mathbf{N}^- \oplus \mathbf{N}^+,$$

as in Theorem 6.2. *Then for each complete Lagrangian subspace* $\mathsf{L} \subset \mathsf{S}$, *the corresponding self-adjoint operator* $\mathbf{T}$ *on* $\mathbf{D}(\mathbf{T})$, *extending* $\mathbf{T}_0$ *on* $\mathbf{D}(\mathbf{T}_0) \subset \mathbf{H}$, *is defined by*

$$\mathsf{L} := \mathbf{D}(\mathbf{T})/\mathbf{D}(\mathbf{T}_0) \text{ or equivalently } \mathbf{D}(\mathbf{T}) = \{\mathbf{f} \in \mathbf{D}(\mathbf{T}_1) : [\mathsf{f} : \mathsf{L}] = 0\},$$

recalling that $\mathsf{f} = \Psi\mathbf{f} = \{\mathbf{f} + \mathbf{D}(\mathbf{T}_0)\}$. *Thus the domain* $\mathbf{D}(\mathbf{T})$ *can also be specified as*

$$\mathbf{D}(\mathbf{T}) = \sum_{k\in\Omega_0} c_k\mathbf{h}_k + \mathbf{D}(\mathbf{T}_0),$$

(with convergence in the $\mathbf{D}$*-norm) where* $\{\mathsf{h}_k : k \in \Omega_0\}$ *is any orthonormal basis for* L *(and hence is a maximal GKN-set in* S*), the set* $\{\mathbf{h}_k \in \mathbf{N}^- \oplus \mathbf{N}^+ \subset \mathbf{D}(\mathbf{T}_1) : k \in \Omega_0\}$ *is determined so that* $\mathsf{h}_k = \Psi\mathbf{h}_k = \{\mathbf{h}_k + \mathbf{D}(\mathbf{T}_0)\}$ *for all* $k \in \Omega_0$*, and the complex numbers* $\{c_k : k \in \Omega_0\}$ *satisfy* $\sum_{k\in\Omega_0} |c_k|^2 < \infty$.

THEOREM 6.6. *Let* $\{I_r, M_r, w_r : r \in \Omega\}$ *be a multi-interval system, with boundary space*

$$\mathsf{S} = \mathbf{D}(\mathbf{T}_1)/\mathbf{D}(\mathbf{T}_0) \approx \mathbf{N}^- \oplus \mathbf{N}^+,$$

where we identify $\mathbf{N}^- \oplus \mathbf{N}^+$ *with* S *as in Theorem* 6.1. *Hence* S *is a complex symplectic space with, for all* $\mathsf{u}, \mathsf{v} \in \mathsf{S}$,

$$[\mathsf{u} : \mathsf{v}] = [\mathbf{u}_- : \mathbf{v}_-]_{\mathbf{D}} + [\mathbf{u}_+ : \mathbf{v}_+]_{\mathbf{D}},$$

and also S *is a Hilbert space with scalar product*

$$\langle \mathsf{u}, \mathsf{v}\rangle_{\mathsf{S}} = \langle \mathbf{u}_-, \mathbf{v}_-\rangle_{\mathbf{D}} + \langle \mathbf{u}_+, \mathbf{v}_+\rangle_{\mathbf{D}}$$

for all $\mathsf{u} \approx \mathbf{u}_- + \mathbf{u}_+, \mathsf{v} \approx \mathbf{v}_- + \mathbf{v}_+$ *in* S*, as in Theorem* 6.5 *above.*

Let $\{\mathsf{b}_k : k \in \Omega_0\}$ *be a (locally) linearly independent set of vectors in* S*, and assume* $\operatorname{span}\{\mathsf{b}_k : k \in \Omega_0\} = \mathsf{L}_0$ *is a Lagrangian submanifold of* S*, in accord with conditions (i) and (ii) of Definition* 6.5*; that is,* $\{\mathsf{b}_k : k \in \Omega_0\}$ *is a GKN-set.*

Then

$$\mathsf{L} := \overline{\mathsf{L}_0} = \overline{\operatorname{span}}\{\mathsf{b}_k : k \in \Omega_0\}$$

is a complete Lagrangian subspace of S *if and only if* $\{\mathsf{b}_k : k \in \Omega_0\}$ *is a maximal GKN-set in* S *(that is, the maximal condition (iii) also holds).*

PROOF. *Let* $\{\mathsf{b}_k : k \in \Omega_0\}$ satisfy conditions (*i*) and (*ii*) of Definition 6.5, and consider the Lagrangian submanifold $\mathsf{L} := \overline{\operatorname{span}}\{\mathsf{b}_k : k \in \Omega_0\} = \overline{\mathsf{L}_0}$ in S.

Assume that $\overline{\mathsf{L}_0} = \mathsf{L}$ is a complete Lagrangian in S. Now take any vector $\mathsf{b} \in \mathsf{S}$ with $b \in \mathsf{L}_0^\perp$ and $[\mathsf{b} : \mathsf{L}] = 0$. We must verify that $\mathsf{b} = 0$ to prove the maximality condition (*iii*). By continuity $\mathsf{b} \in \mathsf{L}^\perp$ and $[\mathsf{b} : \mathsf{L}] = 0$; but L is complete, so $\mathsf{b} \in \mathsf{L}$, and since $\mathsf{L} \cap \mathsf{L}^\perp = 0$ we find that $\mathsf{b} = 0$ as required.

For the converse argument let $\{\mathsf{b}_k : k \in \Omega_0\}$ be a maximal GKN-set in S, and define $\mathsf{L} := \overline{\operatorname{span}}\{\mathsf{b}_k : k \in \Omega_0\} = \overline{\mathsf{L}_0}$ in S. It is clear that L is a Lagrangian closed subspace of S. Now take any vector $\mathsf{b} \in \mathsf{S}$ with $[\mathsf{b} : \mathsf{L}] = 0$. From the orthogonal direct sum decomposition $\mathsf{S} = \mathsf{L} \oplus \mathsf{L}^\perp$ we obtain the corresponding decomposition $\mathsf{b} = \mathsf{b}_{\mathsf{L}} + \mathsf{b}_{\mathsf{L}^\perp}$ in S. Since $[\mathsf{b}_{\mathsf{L}} : \mathsf{L}] = 0$, we find that $[\mathsf{b}_{\mathsf{L}^\perp} : \mathsf{L}] = 0$. Hence

$$[\mathsf{b}_{\mathsf{L}^\perp} : \mathsf{L}_0] = 0 \text{ and } \mathsf{b}_{\mathsf{L}^\perp} \in \mathsf{L}_0^\perp = \mathsf{L}^\perp.$$

The maximality condition then assures that $\mathsf{b}_{\mathsf{L}^\perp} = 0$. Thus $\mathsf{b} = \mathsf{b}_{\mathsf{L}} \in \mathsf{L}$ and we conclude that L is a complete Lagrangian subspace in S. □

COROLLARY 6.5. *Under the hypothesis of Theorem* 6.6 *take any orthonormal set* $\{\mathsf{h}_k : k \in \Omega_0\}$ *in the Hilbert space* S*, and assume* $\mathsf{L}_0 := \operatorname{span}\{\mathsf{h}_k : k \in \Omega_0\}$ *is a Lagrangian submanifold of* S.

Then $\{\mathsf{h}_k : k \in \Omega_0\}$ *is a maximal GKN-set in* S *if and only if* $\mathsf{L} := \overline{\mathsf{L}_0}$ *is a complete Lagrangian subspace of* S. *In this case* $\{\mathsf{h}_k : k \in \Omega_0\}$ *is a complete orthonormal basis for* L *and hence*

$$\mathsf{L} = \sum_{k\in\Omega_0} c_k \mathsf{h}_k \text{ with } \sum_{k\in\Omega_0} |c_k|^2 < \infty,$$

where the convergence is in the Hilbert space norm in L, *and card*$(\Omega_0) = \dim(\mathsf{L})$.

7. Finite multi-interval systems

In the preceding sections of this paper we have investigated multi-interval systems

$$\{I_r, M_r, w_r : r \in \Omega\} \tag{7.1}$$

and the self-adjoint operators $\mathbf{T}$ which they generate in the system Hilbert space $\mathbf{H} = \sum_{r\in\Omega} \oplus L_r^2$, see Definitions 2.1 and 5.1 above. In particular the GKN-Theorem 6.2 determines all such self-adjoint operators in terms of the complete Lagrangian subspaces of the corresponding boundary space

$$\mathsf{S} = \mathbf{D}(\mathbf{T}_1)/\mathbf{D}(\mathbf{T}_0), \tag{7.2}$$

which is a complex symplectic space (in terms of the maximal and minimal operators $\mathbf{T}_1$ on $\mathbf{D}(\mathbf{T}_1)$ and $\mathbf{T}_0$ on $\mathbf{D}(\mathbf{T}_0)$ in $\mathbf{H}$), with a symplectic product $[\cdot : \cdot]_{\mathsf{S}}$ and a Hilbert space scalar product $\langle \cdot, \cdot \rangle_{\mathsf{S}}$ as described in Section 6, especially Definition 6.1 and Theorems 6.1 and 6.3.

Moreover it is there observed that, for the case when Ω is a finite set, say $\{1, 2, \ldots, N\}$ for some positive integer N, the boundary space has the symplectic orthogonal direct sum decomposition (see Definition 3.2)

$$\mathsf{S} = \mathsf{S}_1 \oplus \mathsf{S}_2 \oplus \cdots \oplus \mathsf{S}_N; \tag{7.3}$$

that is $\mathsf{S} = \operatorname{span}\{\mathsf{S}_1, \mathsf{S}_2, \ldots, \mathsf{S}_N\}$ and $\mathsf{S}_r \cap \mathsf{S}_t = 0$ with $[\mathsf{S}_r : \mathsf{S}_t] = 0$ for all $1 \le r \ne t \le N$. Here each subspace S_r is the complex symplectic space

$$\mathsf{S}_r = D(T_{1,r})/D(T_{0,r}) \quad (r = 1, 2, \ldots, N) \tag{7.4}$$

in terms of the maximal and minimal operators $T_{1,r}$ on $D(T_{1,r})$ and $T_{0,r}$ on $D(T_{0,r})$, respectively, in the complex Hilbert space $L_r^2 = L^2(I_r; w_r)$. In this case the dimension of S is

$$\dim(\mathsf{S}) = \sum_{r=1}^{N} \dim(\mathsf{S}_r) < +\infty \quad (\dim(\mathsf{S}_r) < +\infty), \tag{7.5}$$

and conversely $\dim(\mathsf{S}) < +\infty$ implies that the index set Ω is finite (at least with regard to the non-trivial single-interval systems with $\dim(\mathsf{S}_r) \ge 1$). We shall make this standing hypothesis, namely, $1 \le \dim(\mathsf{S}_r) < +\infty$, and refer to such multi-interval systems as *finite*.

In this light we shall now investigate the complete Lagrangian subspaces $\mathsf{L} \subset \mathsf{S}$ in terms of the invariants of $\mathsf{S}_1, \mathsf{S}_2, \ldots, \mathsf{S}_N$, especially with regard to the properties of separation and coupling of the corresponding boundary conditions amongst the various intervals $I_1, I_2, \ldots, I_N$ of the finite multi-interval system (7.1) with (7.3).

We recall the inter-relations between the symplectic invariants of S_r, namely as in Theorem 3.1 for $r = 1, 2, \ldots, N$,

$$\left\{ \begin{array}{rcl} p_r, q_r \text{ with } \Delta_r & = & \min\{p_r, q_r\} \\ \dim(\mathsf{S}_r) & = & D_r = p_r + q_r > 0 \\ Ex(\mathsf{S}_r) & = & p_r - q_r \end{array} \right. \tag{7.6}$$

and the system invariants of S

$$\left\{ \begin{array}{l} \mathbf{p} = \sum_{r=1}^{N} p_r \quad \mathbf{q} = \sum_{r=1}^{N} q_r \quad \mathbf{\Delta} = \min\{\mathbf{p}, \mathbf{q}\} \\ \dim(\mathsf{S}) = \sum_{r=1}^{N} D_r \quad Ex(\mathsf{S}) = \sum_{r=1}^{N} Ex(\mathsf{S}_r). \end{array} \right. \tag{7.7}$$

Since we are seeking complete Lagrangians $\mathsf{L} \subset \mathsf{S}$, we shall always assume that the excess of S satisfies

$$Ex(\mathsf{S}) = 0, \tag{7.8}$$

so that $\mathbf{p} = \mathbf{q} = \mathbf{\Delta}$, which we usually denote by $\mathbf{d}$. There is , of course, no such restriction on the values of $Ex(\mathsf{S}_r)$, except that their sum totals to zero. Then $\dim(\mathsf{S}) = 2\mathbf{d} = \mathbf{D}$ (say), and a Lagrangian $\mathsf{L} \subset \mathsf{S}$ is complete if and only if $\dim(\mathsf{L}) = \mathbf{d}$.

Remark 7.1. At this stage we do not consider further decompositions of the boundary symplectic spaces S_r, for the given interval I_r, into left and right end-point spaces, as was done in great detail for single-interval systems in the monograph [**9**, Section III.1]. Nevertheless, it will be useful to recall some of the prior results and methods for single-interval systems $\{I, M, w\}$ where $\Omega = \{1\}$, with the corresponding boundary complex symplectic space

$$\mathsf{S} = \mathsf{S}_- \oplus \mathsf{S}_+ \text{ with } [\mathsf{S}_- : \mathsf{S}_+] = 0, \tag{7.9}$$

decomposed into the symplectic orthogonal direct sum of the left and right end-point spaces S_- and S_+, with the corresponding invariants $p_\pm, q_\pm, \Delta_\pm$. Since our subsequent treatment of the complex symplectic space

$$\mathsf{S} = \mathsf{S}_1 \oplus \mathsf{S}_2 \oplus \cdots \oplus \mathsf{S}_N$$

will be purely algebraic, these earlier studies can be considered to deal with the case where $N = 2$ (that is as in the earlier case but with S_1 replacing S_- and S_2 replacing S_+).

We shall assume that $Ex(\mathsf{S}) = 0$ so $\mathbf{D} = 2\mathbf{d}$, and a Lagrangian subspace $\mathsf{L} \subset \mathsf{S} = \mathsf{S}_- \oplus \mathsf{S}_+$ is complete just in case $\dim(\mathsf{L}) = \mathbf{d}$.

In this situation the "Balanced Intersection Principle" applies, see [**9**, Section III.1, Theorem 3], for $N = 2$ as in (7.9) (compare with Corollary 7.1 below for $N = 3$),

$$\mathbf{\Delta}_- - \dim(\mathsf{L} \cap \mathsf{S}_-) = \mathbf{\Delta}_+ - \dim(\mathsf{L} \cap \mathsf{S}_+). \tag{7.10}$$

Accordingly we define the *coupling grade* of L, denoted by grade(L), to be this common value in (7.10). Furthermore it is known that there exists a complete Lagrangian for each assigned coupling grade

$$0 \leq \text{grade}(\mathsf{L}) \leq \min\{\mathbf{\Delta}_-, \mathbf{\Delta}_+\}. \tag{7.11}$$

In particular, the complete Lagrangian L is called *strictly separated* in case:

$$\mathbf{d} = \dim(\mathsf{L} \cap \mathsf{S}_-) + \dim(\mathsf{L} \cap \mathsf{S}_+)$$

and this is equivalent to the conditions that grade$(\mathsf{L}) = 0$ and $Ex(\mathsf{S}_\pm) = 0$. This situation happens just in case L has a basis of $\mathbf{d}$ vectors, each of which lies either in the subspace S_- or in the subspace S_+, that is, each basis vector lies in $\mathsf{S}_- \cup \mathsf{S}_+$ (note this is not the same as span$\{\mathsf{S}_-, \mathsf{S}_+\}$).

Also the complete Lagrangian is called *totally coupled* in case:

$$\mathsf{L} \cap \mathsf{S}_- = \mathsf{L} \cap \mathsf{S}_+ = 0$$

and this is equivalent to the conditions grade$(\mathsf{L}) = \min\{\mathbf{\Delta}_-, \mathbf{\Delta}_+\}$ and $\mathbf{\Delta}_- = \mathbf{\Delta}_+$. This happens just in case every basis of L consists of vectors, each of which is coupled at the ends of the interval (that is, no basis vector lies in $\mathsf{S}_- \cup \mathsf{S}_+$).

We shall apply these concepts, suitably modified, to the algebra of finite multi-interval systems, with $\Omega = \{1, 2, \ldots, N\}$ for $N \geq 3$.

Definition 7.1. Consider a complex symplectic space

$$\mathsf{S} = \mathsf{S}_1 \oplus \mathsf{S}_2 \oplus \cdots \oplus \mathsf{S}_N, \tag{7.12}$$

as a finite symplectic orthogonal direct sum of $N \geq 2$ symplectic subspaces with $\dim(\mathsf{S}_r) \geq 1$ for $r = 1, 2, \ldots, N$. Assume

$$\dim(\mathsf{S}) = 2\mathbf{d} < +\infty \text{ and } Ex(\mathsf{S}) = 0.$$

Then a complete Lagrangian subspace $\mathsf{L} \subset \mathsf{S}$ is:

1-separated in case there exists a basis for L consisting of $\mathbf{d}$ vectors, each lying in

$$\bigcup_{r=1}^{N} \mathsf{S}_r$$

(that is, each basis vector belongs to exactly one of the spaces $\{\mathsf{S}_r : r = 1, 2, \ldots, N\}$, recalling that $\mathsf{S}_r \cap \mathsf{S}_t = 0$ for $1 \leq r \neq t \leq N$).

2-separated in case there is a basis for L consisting of $\mathbf{d}$ vectors, each lying in

$$\bigcup_{1 \leq r \neq t \leq N} (\mathsf{S}_r \oplus \mathsf{S}_t)$$

(that is, for each basis vector v we have $v \in \mathsf{S}_r$ for some r, or $v \in$ span$\{\mathsf{S}_r, \mathsf{S}_t\}$ for some r, t, depending on the choice of v, with $1 \leq r \neq t \leq N$; noting again that $\mathsf{S}_r \cap \mathsf{S}_t = 0$ for $1 \leq r \neq t \leq N$ we can state this condition more succinctly as $v \in \bigcup_{1 \leq r \neq t \leq N} \text{span}\{\mathsf{S}_r, \mathsf{S}_t\}$).

In general for any positive integer k we define $\mathsf{L} \subset \mathsf{S}$ to be:

k-separated (with $1 \leq k \leq N - 1$) in case there is a basis for L consisting of $\mathbf{d}$ vectors, each lying in

$$\bigcup_{1 \leq r_1 < r_2 < \ldots < r_k \leq N} (\mathsf{S}_{r_1} \oplus \mathsf{S}_{r_2} \oplus \cdots \oplus \mathsf{S}_{r_k})$$

(that is, each basis vector $v \in \bigcup \text{span}\{\mathsf{S}_{r_1}, \mathsf{S}_{r_2}, \cdots, \mathsf{S}_{r_k}\}$ where this union is taken over all possible k-tuples of distinct indices and corresponding summands of the set $\{\mathsf{S}_1, \mathsf{S}_2, \ldots, \mathsf{S}_N\}$).

Remark 7.2. We note that Definition 7.1 is phrased so that 1-separation of $\mathsf{L} \subset \mathsf{S}$ implies 2-separation, 2-separation implies 3-separation and k-separation for $k > 2$, *etc.* Also note that the property of k-separation for $\mathsf{L} \subset \mathsf{S}$ refers implicitly to the specified direct sum decomposition of S, as prescribed in (7.12) - hence these particular subspaces $\mathsf{S}_1, \mathsf{S}_2, \ldots, \mathsf{S}_N$ are distinguished in Definition 7.1.

In another terminology, compare with [**9**, Section III.1], L is *strictly separated* if and only if L is 1-separated. Further, again with the notation of [**9**, Section, III.1], L is *totally coupled* if and only if each (non-zero) vector $v \in \mathsf{L}$ does not lie in any $(N-1)$-summand, that is, each vector couples all the N coordinates in the sense that

$$v \neq \bigcup_{1 \leq r_1 < r_2 < \ldots < r_{N-1} \leq N} \left(\mathsf{S}_{r_1} \oplus \mathsf{S}_{r_2} \oplus \cdots \oplus \mathsf{S}_{r_{N-1}}\right)$$

(v has non-zero components in each S_r for $r = 1, 2, \ldots, N$).

The next theorem proves that L is never totally coupled, provided that $N \geq 3$.

THEOREM 7.1. *Consider a complex symplectic space*

$$\mathsf{S} = \mathsf{S}_1 \oplus \mathsf{S}_2 \oplus \cdots \oplus \mathsf{S}_N, \tag{7.13}$$

where $N \geq 3$, *with* $\dim(\mathsf{S}) = 2\mathbf{d}$ *and* $Ex(\mathsf{S}) = 0$, *as in Definition* 7.1 *above. Then no complete Lagrangian* $\mathsf{L} \subset \mathsf{S}$ *is totally coupled.*

PROOF. Let S_1 be a subspace with the smallest dimension amongst the set $\{\mathsf{S}_r : r = 1, 2, \ldots, N\}$. Then $0 < D_1 \leq D_r$ for $r \in \{2, \ldots, N\}$ and so $\dim(\mathsf{S}_1) = D_1 \leq \frac{1}{3}\mathbf{D}$; hence $D_1 \leq \frac{2}{3}\mathbf{d} < \mathbf{d}$. But then compute

$$\dim(\mathsf{L}) + \dim(\mathsf{S}_2 \oplus \cdots \oplus \mathsf{S}_N) = \mathbf{d} + D_2 + \cdots D_N > \dim(\mathsf{S}).$$

Therefore the $\mathbf{d}$-space L must have a non-zero intersection with the subspace $(\mathsf{S}_2 \oplus \cdots \oplus \mathsf{S}_N)$ in S. That is, there exists a non-zero vector $v \in \mathsf{L} \cap \{\mathsf{S}_2 \oplus \cdots \oplus \mathsf{S}_N\}$, which proves that L fails to be totally coupled. □

THEOREM 7.2. *Consider a symplectic space, where* $N \geq 3$,

$$\mathsf{S} = \mathsf{S}_1 \oplus \mathsf{S}_2 \oplus \cdots \oplus \mathsf{S}_N \tag{7.14}$$

with $\dim(\mathsf{S}) = 2\mathbf{d}$ *and* $Ex(\mathsf{S}) = 0$, *as in Definition* 7.1 *above. Then there exists a complete Lagrangian* $\mathsf{L} \subset \mathsf{S}$ *which is strictly separated if and only if* $Ex(\mathsf{S}_r) = 0$ *for all* $r = 1, 2, \ldots, N$. *Furthermore, in this case each subspace* S_r *must contain a non-zero vector of* L.

PROOF. First assume that $Ex(\mathsf{S}_r) = 0$ for all $r = 1, 2, \ldots, N$. Then $\dim(\mathsf{S}_r) = 2d_r > 0$ with $\mathbf{d} = \sum_{r=1}^{N} d_r$. In each space S_r there exists a Lagrangian subspace $\mathsf{L}_r \subset \mathsf{S}_r$, which is complete within the complex symplectic space S_r. Now define the Lagrangian subspace L of S by

$$\mathsf{L} = \mathsf{L}_1 \oplus \mathsf{L}_2 \oplus \cdots \oplus \mathsf{L}_N$$

(or $\mathsf{L} = \operatorname{span}\{\mathsf{L}_1, \mathsf{L}_2, \ldots, \mathsf{L}_N\}$ in S). Clearly

$$\dim(\mathsf{L}) = d_1 + d_2 + \cdots + d_N = \mathbf{d},$$

so that L is a complete Lagrangian in S. Since the union of the bases chosen for all L_r constitutes a basis for L, we conclude that L is 1-separated (*i.e.* strictly separated).

Conversely, assume that there is a complete Lagrangian $\mathsf{L} \subset \mathsf{S}$, so $\dim(\mathsf{L}) = \mathbf{d}$, and furthermore that L is 1-separated (strictly separated). Then there exists a basis $\{\mathsf{v}_1, \mathsf{v}_2, \ldots, \mathsf{v}_{\mathbf{d}}\}$ for L with each of these $\mathbf{d}$ vectors lying within a single one of the given summands $\mathsf{S}_1, \mathsf{S}_2, \ldots, \mathsf{S}_N$ (perhaps different basis vectors in different summands). We show next that each excess $Ex(\mathsf{S}_r) = 0$ for $r = 1, 2, \ldots, N$.

Note that at least one of these basis vectors $\{\mathsf{v}_1, \mathsf{v}_2, \ldots, \mathsf{v}_{\mathbf{d}}\}$ must lie in S_1. Because otherwise there exists a non-zero vector $\mathsf{u}_1 \in \mathsf{S}_1$ (or $\mathsf{u} = (\mathsf{u}_1, 0, 0, \ldots, 0) \in \mathsf{S}$) with $[\mathsf{u}_1 : \mathsf{L}] = 0$ yet $\mathsf{u}_1 \notin \mathsf{L}$, which contradicts the assumption that L is a complete Lagrangian in S. Hence we can assume that each summand S_r, for $r = 1, 2, \ldots, N$, must contain at least one of the basis vectors $\{\mathsf{v}_1, \mathsf{v}_2, \ldots, \mathsf{v}_{\mathbf{d}}\}$.

Define the subsets $\mathsf{L}_r := \mathsf{L} \cap \mathsf{S}_r$ which are non-zero Lagrangians subspaces of S_r, for each $r = 1, 2, \ldots, \mathbf{d}$. We show that each L_r is a complete Lagrangian in S_r.

Take a non-zero vector $\mathsf{u}_1 \in \mathsf{S}_1$ and assume that $[\mathsf{u}_1 : \mathsf{L}_1] = 0$. Since $\mathsf{u}_1 \in \mathsf{S}_1$, we observe that $[\mathsf{u}_1 : \mathsf{S}_r] = 0$ for $r = 2, 3, \ldots, N$. Hence, $[\mathsf{u}_1 : \mathsf{v}_r] = 0$ for each of the given basis vectors $\{\mathsf{v}_1, \mathsf{v}_2, \ldots, \mathsf{v}_{\mathbf{d}}\}$ of L. Therefore $[\mathsf{u}_1 : \mathsf{L}] = 0$ and so $\mathsf{u}_1 \in \mathsf{L}$, and thus $\mathsf{u}_1 \in \mathsf{L} \cap \mathsf{S}_1 = \mathsf{L}_1$. This proves that L_1 is a complete Lagrangian in S_1, and similar results hold for each L_r in S_r for $r = 1, 2, \ldots, N$.

Since each complex symplectic space S_r contains a complete Lagrangian subspace $\mathsf{L}_r \subset \mathsf{S}_r$, we conclude that $Ex(\mathsf{S}_r) = 0$ for each $r = 1, 2, \ldots, N$. □

A weak form of the Balanced Intersection Principle (7.10) applies to the situation in Theorem 7.2 when $N = 3$, but say $Ex(\mathsf{S}_1) \neq 0$.

COROLLARY 7.1. *Consider a complex symplectic space*

$$\mathsf{S} = \mathsf{S}_1 \oplus \mathsf{S}_2 \oplus \mathsf{S}_3$$

with $\dim(\mathsf{S}) = 2\mathbf{d}$ *and* $Ex(\mathsf{S}) = 0$, *as in Theorem* 7.2 *where* $N = 3$, *and let* L *be a complete Lagrangian subspace of* S.

Assume $Ex(\mathsf{S}_3) = 0$ *and that* $\mathsf{L} \cap \mathsf{S}_3$ *is a complete Lagrangian in* S_3. *Then:*

$$\Delta_1 - \dim(\mathsf{L} \cap \mathsf{S}_1) = \Delta_2 - \dim(\mathsf{L} \cap \mathsf{S}_2).$$

In particular, if further $Ex(\mathsf{S}_2) = 0$ *then* $Ex(\mathsf{S}_1) = 0$, *and in this case:*

$$\mathsf{L} \cap \mathsf{S}_1 \text{ is complete in } \mathsf{S}_1, \text{ if and only if, } \mathsf{L} \cap \mathsf{S}_2 \text{ is complete in } \mathsf{S}_2.$$

Moreover if both $\mathsf{L} \cap \mathsf{S}_1$ *and* $\mathsf{L} \cap \mathsf{S}_2$ *are so complete, then* L *is strictly separated in* S.

PROOF. Since $Ex(\mathsf{S}_3) = 0$, then $Ex(\mathsf{S}_1 \oplus \mathsf{S}_2) = 0$. We can apply the Balanced Intersection Principle to the symplectic space S with the two direct summands $(\mathsf{S}_1 \oplus \mathsf{S}_2)$ and S_3 to obtain (recalling that $p_1 + p_2 = q_1 + q_2 = \Delta(\mathsf{S}_1 \oplus \mathsf{S}_2)$ and $p_3 = q_3 = \Delta_3$)

$$p_1 + p_2 - \dim(\mathsf{L} \cap (\mathsf{S}_1 \oplus \mathsf{S}_2)) = \Delta_3 - \dim(\mathsf{L} \cap \mathbb{S}_3).$$

But $\mathsf{L} \cap \mathsf{S}_3$ is complete in S_3, so $\dim(\mathsf{L} \cap \mathsf{S}_3) = \Delta_3 = \frac{1}{2}\dim(\mathsf{S}_3)$. Hence $\dim(\mathsf{L} \cap (\mathsf{S}_1 \oplus \mathsf{S}_2)) = p_1 + p_2 = \frac{1}{2}\dim(\mathsf{S}_1 \oplus \mathsf{S}_2)$, so $\mathsf{L} \cap (\mathsf{S}_1 \oplus \mathsf{S}_2)$ is complete in the symplectic space $\mathsf{S}_1 \oplus \mathsf{S}_2$. Again an application of (7.10) to the space $\mathsf{S}_1 \oplus \mathsf{S}_2$ yields

$$\Delta_1 - \dim(\mathsf{L} \cap \mathsf{S}_1) = \Delta_2 - \dim(\mathsf{L} \cap \mathsf{S}_2).$$

Finally if we further assume that $Ex(\mathsf{S}_2) = 0$, so then $Ex(\mathsf{S}_1) = 0$, we note that $\Delta_1 = \frac{1}{2}\dim(\mathsf{S}_1)$ and $\Delta_2 = \frac{1}{2}\dim(\mathsf{S}_2)$. Thus the conclusion of the Corollary follows. □

Remark 7.3. Consider a finite multi-interval system $\{I_r, M_r, w_r : r \in \Omega\}$ where $\Omega = \{1, 2, \ldots, N\}$. Assume that each interval I_r is compact, so that the corresponding boundary value problem is regular, and hence S_r has excess $Ex(\mathsf{S}_r) = 0$. In this case there must exist a complete Lagrangian $\mathsf{L} \subset \mathsf{S}$ which is strictly separated, that is, there exists a basis for L defining self-adjoint boundary conditions for the multi-interval system, and such that each boundary condition involves data only at the endpoints of just a single interval I_r. Also each interval is involved in at least one of the boundary conditions specifying L.

THEOREM 7.3. *Consider a complex symplectic space, where* $N \geq 3$,

$$\mathsf{S} = \mathsf{S}_1 \oplus \mathsf{S}_2 \oplus \cdots \oplus \mathsf{S}_N \tag{7.15}$$

with $\dim(\mathsf{S}) = 2\mathbf{d} < +\infty$ *and* $Ex(\mathsf{S}) = 0$, *as in Definition* 7.1 *above. Then there exists a complete Lagrangian* $\mathsf{L} \subset \mathsf{S}$ *which is 2-separated.*

PROOF. We consider a basis of $2\mathbf{d}$ vectors for S consisting of $\dim(\mathsf{S}_r) = 2d_r$ vectors in each symplectic space S_r, and chosen so that the corresponding skew-hermitian matrix for S_r is diagonalized as in [**9**, Section III.1, Theorem 1], for $r = 1, 2, \ldots, N$,

$$\begin{bmatrix} -iI_{q_r} & 0 \\ 0 & iI_{p_r} \end{bmatrix}.$$

Denote these basis vectors as, for S_r with $r = 1, 2, \ldots, N$,

$$\{e_1^r, e_2^r, \ldots, e_{q_r}^r, f_1^r, f_2^r, \ldots, f_{p_r}^r\}$$

with, for $1 \leq k_1, k_2 \leq q_r$ and $1 \leq l_1, l_2 \leq p_r$,

$$[e_{k_1}^r : e_{k_2}^r] = -i\delta_{k_1 k_2} \qquad [f_{l_1}^r : f_{l_2}^r] = i\delta_{l_1 l_2}.$$

Furthermore these $2\mathbf{d}$ vectors are pairwise symplectically orthogonal in S.

Now there is a total of $\mathbf{d}$ vectors in the set

$$\mathbf{E} := \{e_1^1, e_2^1, \ldots, e_{q_1}^1, e_1^2, e_2^2, \ldots, e_{q_2}^2, \ldots, e_1^N, e_2^N, \ldots, e_{q_N}^N\} \tag{7.16}$$

and also $\mathbf{d}$ vectors in the set

$$\mathbf{F} := \{f_1^1, f_2^1, \ldots, f_{p_1}^1, f_1^2, f_2^2, \ldots, f_{p_2}^2, \ldots, f_1^N, f_2^N, \ldots, f_{p_N}^N\} \tag{7.17}$$

since

$$Ex(\mathsf{S}) = \sum_{r=1}^{N} p_r - \sum_{r=1}^{N} q_r = 0$$

and

$$\dim(\mathsf{S}) = \sum_{r=1}^{N} p_r + \sum_{r=1}^{N} q_r = 2\mathbf{d}.$$

Now choose a one-to-one correspondence G between these two sets (7.16) and (7.17) of $\mathbf{d}$ vectors

$$G : \mathbf{E} \to \mathbf{F}$$

and denote this re-arrangement $\mathbf{\Phi}$ of $\mathbf{F}$ by

$$\mathbf{\Phi} := \{\varphi_k^r := G(e_k^r) \text{ for } 1 \leq k \leq q_r \text{ and } 1 \leq r \leq N\},$$

so each φ_k^r is one of the vectors in the set $\mathbf{F}$. Now define $\mathbf{d}$ vectors in S by

$$\{e_k^r + \varphi_k^r : 1 \leq k \leq q_r \text{ and } 1 \leq r \leq N\}. \tag{7.18}$$

Each of the $\mathbf{d}$ vectors in the set (7.18) depends on exactly one of the vectors $\{e_1^1, \ldots, e_{q_N}^N\}$, and also on exactly one of the corresponding vectors $\{f_1^1, \ldots, f_{p_N}^N\}$ since the set $\mathbf{\Phi}$ is a re-arrangement of the set $\mathbf{F}$. Further these $\mathbf{d}$ vectors of (7.18), just as the vectors of (7.16), are linearly independent in S. Moreover each member of this set (7.18) is a neutral vector in the sense

$$[e_k^r + \varphi_k^r : e_k^r + \varphi_k^r] = -i + i = 0 \text{ for all } 1 \le k \le q_r \text{ and } 1 \le r \le N.$$

Thus this set of vectors (7.18) constitutes a basis for a Lagrangian subspace $\mathsf{L} \subset \mathsf{S}$, and since $\dim(\mathsf{L}) = \mathbf{d}$ it follows that L is a complete Lagrangian space.

But each vector $e_k^r + \varphi_k^r$ lies either in one summand, in the case when $\varphi_k^r = G(e_k^r) \in \mathsf{S}_r$, or else it lies in $\operatorname{span}\{\mathsf{S}_r, \mathsf{S}_t\}$ when $\varphi_k^r = G(e_k^r) \in \mathsf{S}_t$ with $t \ne r$. Thus each basis vector of the $\mathbf{d}$ vectors in (7.18) lies within the span of at most two of the summand spaces $\{\mathsf{S}_r : r = 1, 2, \cdots, N\}$ and therefore L is 2-separated in S. □

We close this discussion of the finite multi-interval systems, and the algebra of the corresponding complex symplectic spaces, with some interesting examples.

Example 7.1 Consider the symplectic space

$$\mathsf{S} = \mathsf{S}_1 \oplus \mathsf{S}_2 \oplus \mathsf{S}_3$$

with $Ex(\mathsf{S}) = 0$ and $\dim(\mathsf{S}) = 6$; here each $\dim(\mathsf{S}_r) = 2$ and $Ex(\mathsf{S}_r) = 0$ for $r = 1, 2, 3$. We construct a complete Lagrangian $\mathsf{L} \subset \mathsf{S}$ which is not 2-separated.

Let us take bases for $\mathsf{S}_1, \mathsf{S}_2, \mathsf{S}_3$

$$\{e_+^1, e_-^1\} \qquad \{e_+^2, e_-^2\} \qquad \{e_+^3, e_-^3\},$$

respectively, with the symplectic products in each S_r for $r = 1, 2, 3$ given by

$$[e_+^r : e_+^r] = i \qquad [e_-^r : e_-^r] = -i \qquad [e_+^r : e_-^r] = 0.$$

Define the Lagrangian 3-space $\mathsf{L} := \operatorname{span}\{u, v, w\} \subset \mathsf{S}$, by the choice of basis vectors

$$\begin{aligned} u &= (e_+^1 + e_-^1) & &+ \tfrac{1}{\sqrt{2}}(e_+^3 + e_-^3) \\ v &= & (e_+^2 + e_-^2) &+ \tfrac{1}{\sqrt{2}}(e_+^3 + e_-^3) \\ w &= \quad e_-^1 & + e_-^2 & + \sqrt{2} e_+^3. \end{aligned}$$

It is straightforward to verify that u, v, w are independent neutral vectors and symplectically orthogonal in S, and hence span a complete Lagrangian $\mathsf{L} \subset \mathsf{S}$.

While u and v each have non-zero components in just two summands (i.e. are 2-coupled),the vector w is more than 2-coupled; however we must show that each basis of L contains a vector which is more than 2-coupled.

Take any other basis of three vectors in L; one of these vectors must depend upon w, say $au + bv + cw$ with $c \ne 0$. Then

$$\begin{aligned} au + bv + cw &= ae_+^1 + (a+c)e_-^1 + be_+^2 + (b+c)e_-^2 \\ &\quad + \left\{a(e_+^3 + e_-^3) + b(e_+^3 + e_-^3) + 2ce_+^3\right\}/\sqrt{2}. \end{aligned}$$

But this vector has a non-zero component in S_1 (namely: $\{a, a+c\}$), also a non-zero component in S_2 (namely: $\{b, b+c\}$), and its component in S_3 is $\{a + b + 2c, a + b\}/\sqrt{2} \ne 0$ (because if $a + b = 0$ then $\{c, 0\} \ne 0$ in S_3).

Hence the complete Lagrangian L is not 2-separated in S.

In Theorem 7.3 we showed that a complex symplectic space S, with a finite decomposition

$$\mathsf{S} = \mathsf{S}_1 \oplus \mathsf{S}_2 \oplus \cdots \oplus \mathsf{S}_N$$

where $\dim(\mathsf{S}) < +\infty$, always contains a complete Lagrangian $\mathsf{L} \subset \mathsf{S}$ (provided that $Ex(\mathsf{S}) = 0$) which is 2-separated. However in Theorem 7.2 we noted that only in very special circumstances ($Ex(\mathsf{S}_r) = 0$ for $r = 1, 2, \ldots, N$) does there exist a 1-separated (equivalently, strictly separated) complete Lagrangian in S.

Further the preceding Example 7.1 exhibits the difficulty in constructing a complete Lagrangian which is not 2-separated in S. The next example, Example 7.2, presents a construction for complete Lagrangians which are not k-separated (hence not 1-separated, 2-separated, *etc.*) for arbitrarily large integers k, in suitable complex symplectic spaces.

In particular, for each positive integer $d \geq 2$ consider the complex symplectic space

$$\mathsf{S} = \mathsf{S}_1 \oplus \mathsf{S}_2 \oplus \cdots \oplus \mathsf{S}_d,$$

where each summand S_r has $\dim(\mathsf{S}_r) = 2$, $Ex(\mathsf{S}_r) = 0$ for $r = 1, 2, \ldots, d$; thus $\dim(\mathsf{S}) = 2d$ and $Ex(\mathsf{S}) = 0$. These invariants define S uniquely, up to symplectic isomorphism. Moreover, in the subsequent discussion referring to the properties of separation, the symplectically orthogonal direct summands S_r are relevant and distinguished.

We construct a complete Lagrangian $\mathsf{L}_d \subset \mathsf{S}$, with $\dim(\mathsf{L}_d) = d$, which is not $[d/2]$-separated; that is each basis for L_d contains a vector which has non-zero components in more than $[d/2]$ of the summands $\mathsf{S}_1, \mathsf{S}_2, \ldots, \mathsf{S}_d$ or, equivalently the vector has a non-zero projection in more than $[d/2]$ of these summands. (Here the symbol $[d/2]$ refers to the "integer part" of $d/2$, so that $[d/2] = d/2$ when d is even, and $[d/2] = (d-1)/2$ when d is odd.)

Remark 7.4. As an application of the proposed construction of $\mathsf{L}_d \subset \mathsf{S}$ we can consider the multi-interval system $\{I_r, M_r, w_r : r = 1, 2, \ldots, d\}$, where for all $r = 1, 2, \ldots, d$:

(1) The interval I_r is compact.
(2) M_r is the first-order classical differential expression, see Section 2 above, given by $M_r[y] := iy'$
(3) The weight $w_r = 1$ on I_r.

Then each single-interval system $\{I_r, M_r, w_r\}$ is regular in $L_r^2 \equiv L^2(I_r; w_r)$ and the deficiency indices $d_r^{\pm} = 1$; for the multi-interval system in $\mathbf{H} = \sum_{r=1}^{d} \oplus L_r^2$ the deficiency indices $\mathbf{d}^{\pm} = d$.

Then the construction given below implies that there must exist a self-adjoint operator $\mathbf{T}$ on $\mathbf{D}(\mathbf{T}) \subset \mathbf{H}$ generated by this system, such that the domain $\mathbf{D}(\mathbf{T})$ cannot be defined by the (required) number d of boundary conditions each of which involves at most $[d/2]$ of the given compact intervals $\{I_r : r = 1, 2, \ldots, d\}$.

Before proceeding with the difficult constructions for Example 7.2, we first illustrate these novel methods for the case $d = 2$. Hence we consider the complex symplectic 4-space $\mathsf{S} = \mathsf{S}_1 \oplus \mathsf{S}_2$ with $\dim(\mathsf{S}_1) = \dim(\mathsf{S}_2) = 2$, $Ex(\mathsf{S}_1) = Ex(\mathsf{S}_2) = 0$.

Fix bases $\{e_-^1, e_+^1\}$ for S_1 and $\{e_-^2, e_+^2\}$ for S_2 with the symplectic products given by

$$[e_-^r : e_-^r] = -i, \quad [e_-^r : e_+^r] = 0, \quad [e_+^r : e_+^r] = +i \text{ for } r = 1, 2, \text{ and } [e_\pm^1 : e_\pm^2] = 0.$$

Clearly the set $\{e_-^1, e_+^1, e_-^2, e_+^2\}$ constitutes a basis for S.

It is straightforward to show that the Lagrangian subspace

$$\mathsf{L}_1 := \operatorname{span}\{e_-^1 + e_+^1, e_-^2 + e_+^2\} \tag{7.19}$$

is complete ($\dim(\mathsf{L}_1) = 2$); but L_1 is 1-separated, and is not the Lagrangian that we seek.

Continuing, following the methods of the earlier Example 3.1 of Section 3 above, we now define two symplectic subspaces H_- and H_+ of S,

$$H_- := \operatorname{span}\{e_-^1, e_-^2\} \text{ and } H_+ := \operatorname{span}\{e_+^1, e_+^2\}.$$

Then we can introduce a different decomposition of S,

$$\mathsf{S} = H_- \oplus H_+,$$

so that each vector $\mathsf{u} \in \mathsf{S}$ has the unique representation $\mathsf{u} = \{u_-, u_+\}$, with $u_\pm \in H_\pm$, respectively. Moreover, on H_- we define the complex hermitian scalar product

$$\langle u_-, v_- \rangle_- := i[u_- : v_-],$$

so that $\{e_-^1, e_-^2\}$ forms an orthonormal basis for H_-; and similarly on H_+ the scalar product

$$\langle u_+, v_+ \rangle_+ := -i[u_+ : v_+]$$

yields a Hilbert space with orthonormal basis $\{e_+^1, e_+^2\}$. Hence for vectors $\mathsf{u} := \{u_-, u_+\}$ and $\mathsf{v} := \{v_-, v_+\}$ in S, see (3.9),

$$[\mathsf{u} : \mathsf{v}]_{\mathsf{S}} = -i\langle u_-, v_- \rangle_- + i\langle u_+, v_+ \rangle_+$$

and S is also a Hilbert space with scalar product

$$\langle \mathsf{u}, \mathsf{v} \rangle_{\mathsf{S}} := \langle u_-, v_- \rangle_- + \langle u_+, v_+ \rangle_+,$$

as in Section 6 above.

Further it is known [**9**, Section III.1, Lemma 1, Page 34] that every complete Lagrangian of S can be represented (uniquely) in the format graph(U) for a unitary surjection U of the Hilbert space H_- onto the Hilbert space H_+ (and each such graph(U) defines a complete Lagrangian in S). In particular the complete Lagrangian L_1, defined by (7.19), is given by $\mathsf{L}_1 = \operatorname{graph}(U_1)$ where the unitary map $U_1 : H_- \to H_+$ is defined by $U_1(e_-^r) := e_+^r$ for $r = 1, 2$.

Now we perform a "rotation" map $A : H_+ \to H_+$, say $u_+ \to \hat{u}_+$ where

$$\hat{u}_+ := Au_+. \tag{7.20}$$

Here the 2×2 matrix A is, for some $\theta \in [0, 2\pi]$,

$$A = \begin{bmatrix} \cos(\theta) & \sin(\theta) \\ -\sin(\theta) & \cos(\theta) \end{bmatrix},$$

and the mapping (7.20) is determined by, in terms of the basis vectors $e_+^1 = [1, 0]^t$ and $e_+^2 = [0, 1]^t$,

$$\begin{aligned} \hat{e}_+^1 &:= \cos(\theta)e_+^1 - \sin(\theta)e_+^2 \\ \hat{e}_+^2 &:= \sin(\theta)e_+^1 + \cos(\theta)e_+^2. \end{aligned}$$

Moreover we choose $\theta \in [0, 2\pi]$ so that the rotation matrix A is in "general position", that is we require $\theta \neq \{0, \frac{1}{2}\pi, \pi, \frac{3}{2}\pi, 2\pi\}$ so that $\cos(\theta) \neq 0$ and $\sin(\theta) \neq 0$ both hold.

Because A is a rotation map of H_+ onto H_+, it is unitary and $\{\hat{e}_+^1, \hat{e}_+^2\}$ constitutes an orthonormal basis for H_+.

We now define the required Lagrangian $\mathsf{L}_2 \subset \mathsf{S}$ by

$$\mathsf{L}_2 := \operatorname{graph}(U_2) \tag{7.21}$$

where the unitary map $U_2 : H_- \to H_+$ is determined by $U_2(e_-^r) := \hat{e}_+^r$ for $r = 1, 2$.

Now we have to show that for each basis $\{\varphi^1, \varphi^2\}$ for L_2 (within the Hilbert space S), at least one of the basis vectors, call it φ, must have non-zero components in both S_1 and S_2; that is, φ does not lie within just one of the spaces S_1 or S_2. This result then verifies that L_2 is 2-separated, but not 1-separated, as required.

For this purpose let φ be any non-zero vector in L_2, so that $\varphi = \varphi_- + \varphi_+$ with $\varphi_\pm \in H_\pm$ and $\varphi_+ = U_2\varphi_-$, using (7.21) above. Now expand $\varphi_\pm$ in the basis $\{e^1_-, e^2_-, e^1_+, e^2_+\}$ of S; that is for complex numbers $a_r \in \mathbb{C}$ with $r = 1, 2$

$$\varphi_- = a_1 e^1_- + a_2 e^2_-$$

and

$$\begin{aligned}\varphi_+ &= a_1 \hat{e}^1_+ + a_2 \hat{e}^2_+ \\ &= a_1(\cos(\theta)e^1_+ - \sin(\theta)e^2_+) + a_2(\sin(\theta)e^1_+ + \cos(\theta)e^2_+).\end{aligned}$$

Not both a_1 and a_2 can vanish, so consider the three cases:

(1) $a_1 \neq 0$ and $a_2 = 0$; then

$$\varphi_- = a_1 e^1_- \text{ and } \varphi_+ = a_1(\cos(\theta)e^1_+ - \sin(\theta)e^2_+)$$

so that φ_+ projects non-trivially into both S_1 and S_2.

(2) $a_1 \neq 0$ and $a_2 \neq 0$; then

$$\varphi_- = a_1 e^1_- + a_2 e^2_-$$

so that φ_- projects non-trivially into both S_1 and S_2.

(3) $a_1 = 0$ and $a_2 \neq 0$; then

$$\varphi_- = a_2 e^2_- \text{ and } \varphi_+ = a_2(\sin(\theta)e^1_+ + \cos(\theta)e^2_+)$$

so that φ_+ projects non-trivially into both S_1 and S_2.

Therefore, in every case L_2 has a basis vector that fails to lie only within S_1, and fails to lie only within S_2; thus L_2 is not 1-separated and the demonstration is finished for the case $d = 2$.

In order to extend these methods to the case when $d > 2$ we utilize rotation matrices $A \in SO(d, \mathbb{R})$ (the special group of all $d \times d$ real orthogonal matrices with $\det(A) = +1$), which are in "general position", as detailed next.

Definition 7.2. For each integer $d \geq 2$ consider the rotation group $SO(d, \mathbb{R})$ of all $d \times d$ real orthogonal matrices $A = [\alpha^i_j]$ with $\det(A) = +1$. Using the d^2 entries of $[\alpha^i_j]$, in some agreed ordering, as coordinates in $\mathbb{R}^{d^2}$, we can consider $SO(d, \mathbb{R})$ as a compact connected subset of this Euclidean space, and hence induce the relative metric topology on $SO(d, \mathbb{R})$.

Define $A = [\alpha^i_j] \in SO(d, \mathbb{R})$ to be in "general position" is case:

(1) $\alpha^i_j \neq 0$ for all $1 \leq i, j \leq d$, and also
(2) each $l \times l$ subdeterminant of A (upon specifying any l rows and l columns) is also non-zero, for $1 \leq l \leq d$.

Note that if A is in general position, then so is its transpose A^t and also its inverse $A^{-1} = A^t$.

PROPOSITION 7.1. *For each integer $d \geq 2$ the set of all rotation matrices in general position constitutes an open dense subset of $SO(d, \mathbb{R})$.*

PROOF. It is known [**25**] that $SO(d,\mathbb{R})$ is a connected real-analytic manifold, of dimension $d(d-1)/2$, in the Euclidean space $\mathbb{R}^{d^2}$ (for instance, $SO(2,\mathbb{R})$ is a circle in $\mathbb{R}^4$). That is, $SO(d,\mathbb{R})$ can be covered by local coordinates charts, each using $d(d-1)/2$ local coordinates defined via projections of some open subset of $SO(d,\mathbb{R})$ onto an open subset of some appropriate coordinate $d(d-1)/2$-plane of $\mathbb{R}^{d^2}$. Moreover on the overlap of two such local charts the corresponding local coordinates are inter-related by real-analytic coordinate transformations, so as to define the real analytic manifold $SO(d,\mathbb{R})$.

Now fix a pair of indices (i,j) with $1 \leq i,j \leq d$ to indicate a matrix entry or position, and define the mapping $(i,j)(\cdot) : SO(d,\mathbb{R}) \to \mathbb{R}$ by $(i,j)(A) = \alpha^i_j$ for all $A \in SO(d,\mathbb{R})$. We shall indicate this function by the notation α^i_j, which also denotes a coordinate function in $\mathbb{R}^{d^2}$. Thus α^i_j indicates a real function, in fact a linear polynomial on $\mathbb{R}^{d^2}$, and this restricts to a real-analytic function on the manifold $SO(d,\mathbb{R})$, which we still denote by the symbol α^i_j. Clearly, $\alpha^i_j = 0$ defines a closed subset of $SO(d,\mathbb{R})$; which consists of all rotation matrices with zero in the (i,j) position. Since α^i_j is a real-analytic function on the real-analytic manifold $SO(d,\mathbb{R})$, it can be expressed as an absolutely convergent real power series, in terms of the local coordinates about any chosen point on $SO(d,\mathbb{R})$. Hence, if $\alpha^i_j \equiv 0$ on some open neighbourhood of a point $A_0 \in SO(d,\mathbb{R})$, then the methods of analytic continuation apply to prove that $\alpha^i_j \equiv 0$ everywhere on $SO(d,\mathbb{R})$. But this result is evidently false, since there exist rotation matrices with $\alpha^i_j = 1$. Thus we conclude that the subset of $SO(d,\mathbb{R})$ that is defined by the condition $\alpha^i_j = 0$ is closed and nowhere dense. The complementary subset $\alpha^i_j \neq 0$ is thus open and dense in $SO(d,\mathbb{R})$.

But there are only a finite number (in fact d^2) entry positions in these rotation matrices, and the intersection of a finite collection of open-dense subsets is itself open and dense in $SO(d,\mathbb{R})$. We conclude that there is an open and dense subset of $SO(d,\mathbb{R})$ consisting of rotation matrices with all entries non-zero, as for condition 1 of Definition 7.2.

Next fix a choice of an $l \times l$ subdeterminant position in the rotation matrices of $SO(d,\mathbb{R})$; say by specifying l rows and l columns. But such an $l \times l$ determinant position is expressed as a polynomial in the coordinate functions α^i_j, and hence restricts to a real-analytic function on $SO(d,\mathbb{R})$. Then the same argument as before shows that the rotation matrices, for which this particular subdeterminant is non-zero, fill an open dense subset of $SO(d,\mathbb{R})$, as for condition 2 of Definition 7.2.

There are only a finite number of such subdeterminant positions, and hence there is an open-dense set of rotation matrices in general position in $SO(d,\mathbb{R})$. □

Example 7.2. Let $d \geq 2$ and consider the complex symplectic space

$$\mathsf{S} = \mathsf{S}_1 \oplus \mathsf{S}_2 \oplus \cdots \oplus \mathsf{S}_d,$$

where the direct summands S_r are specified with

$$\dim(\mathsf{S}_r) = 2 \text{ and } Ex(\mathsf{S}_r) = 0 \text{ for } r = 1, 2, \ldots, d.$$

Then there exists a complete Lagrangian subspace $\mathsf{L}_d \subset \mathsf{S}$ which is not $[d/2]$-separated.

We construct this example as follows.

For each $r = 1, 2, \ldots, d$ consider the basis vectors $e_{\pm}^r \in \mathsf{S}_r$, with symplectic products

$$[e_-^r : e_-^r] = -i \qquad [e_+^r : e_+^r] = i \qquad [e_-^r : e_+^r] = 0.$$

Also, since any two summands S_r and S_t, with $r \neq t$, are symplectically orthogonal

$$[e_{\pm}^r : e_{\pm}^t] = 0 \text{ for } 1 \leq r \neq t \leq d.$$

Then this set of $2d$ vectors $\{e_{\pm}^r : r = 1, 2, \ldots, d\}$ constitutes a basis for S.

Now define the complex symplectic subspaces $H_{\pm} \subset \mathsf{S}$

$$H_- := \operatorname{span}\{e_-^r : r = 1, 2, \ldots, d\} \text{ and } H_+ := \operatorname{span}\{e_+^r : r = 1, 2, \ldots, d\}.$$

Then $\mathsf{S} = H_- \oplus H_+ is$ a symplectically orthogonal decomposition of S, and each vector $\mathsf{u} \in \mathsf{S}$ has a unique decomposition $\mathsf{u} = \{u_-, u_+\}$ with $u_{\pm} \in H_{\pm}$, respectively. Furthermore, each of $H_{\pm}$ is a Hilbert space with the corresponding scalar products

$$\langle u_-, v_- \rangle_- = i[u_- : v_-] \qquad \langle u_+, v_+ \rangle_+ = -i[u_+ : v_+]$$

and S is also a Hilbert space with the scalar product

$$\langle \mathsf{u}, \mathsf{v} \rangle_{\mathsf{S}} = \langle u_-, v_- \rangle_- + \langle u_+, v_+ \rangle_+$$

so the norms satisfy

$$\|\mathsf{u}\|_{\mathsf{S}}^2 = \|u_-\|_-^2 + \|u_+\|_+^2$$

for all vectors $\mathsf{u} = \{u_-, u_+\}$ and $\mathsf{v} = \{v_-, v_+\}$ in S.

As in the preliminary discussion (where $d = 2$), we have the 1-separated complete Lagrangian

$$\mathsf{L}_1 := \{e_-^1 + e_+^1, \ldots, e_-^d + e_+^d\}$$

and now we introduce a "general rotation" of H_+ in order to define the complete Lagrangian $\mathsf{L}_d \subset \mathsf{S}$. Namely, let $A = [\alpha_r^\sigma]$ be a rotation matrix in general position in $SO(d, \mathbb{R})$ (as in Definition 7.2), and consider the rotation map

$$A : H_+ \to H_+ \text{ with } e_+^r \to \hat{e}_+^r := A e_+^r \text{ for } r = 1, 2, \ldots, d,$$

again defined in terms of the coordinate basis $e_+^1 = [1, 0, \cdots, 0]^t, \ldots,$
$e_+^d = [0, \cdots, 0, 1]^t$ in H_+, *i.e.*

$$\hat{e}_+^r = \sum_{\sigma=1}^d \alpha_r^\sigma e_+^\sigma$$

(with coefficient matrix $A^t = A^{-1}$). Clearly $\{\hat{e}_+^1, \ldots, \hat{e}_+^d\}$ is then an orthonormal basis for the Hilbert space H_+.

Now define the linear map U_d of H_- onto H_+ by $U_d e_-^r = \hat{e}_+^r$ for $r = 1, 2, \ldots, d$. Since U_d carries the orthonormal basis $\{e_-^1, \ldots, e_-^d\}$ of H_- onto an orthonormal basis of H_+, the map U_d is a unitary surjection of H_- onto H_+.

Next define the required complete Lagrangian $\mathsf{L}_d \subset \mathsf{S}$ by

$$\mathsf{L}_d := \operatorname{graph}(U_d); \tag{7.22}$$

that is, $\varphi = \varphi_- + \varphi_+$ lies in L_d if and only if $\varphi_+ = U_d \varphi_-$, with $\varphi_{\pm} \in H_{\pm}$ as before, see [**9**, Section III, Lemma 1, Page 34] We shall verify that L_d is not $[d/2]$-separated.

Let $\varphi = \varphi_- + \varphi_+$ be any non-zero vector in L_d, and expand each term relative to the basis $\{e_-^1, e_+^1, \ldots, e_-^d, e_+^d\}$ of S. Then

$$\varphi_- = \sum_{r=1}^d a_r e_-^r \text{ and } \varphi_+ = \sum_{r=1}^d a_r \hat{e}_r^r,$$

or, equivalently,

$$\varphi_+ = \sum_{r=1}^{d} a_r(\alpha_r^1 e_+^1 + \cdots + \alpha_r^d e_+^d),$$

and not all the complex coefficients $\{a_1, \ldots, a_d\}$ are zero.

Now consider the following cases:

(1) Assume one of the coefficients, say $a_1 \neq 0$, and all the remaining coefficients $a_r = 0$ for $r = 2, 3, \ldots, d$. Then

$$\varphi_- = a_1 e_-^1 \text{ and } \varphi_+ = a_1(\alpha_1^1 e_+^1 + \cdots + \alpha_1^d e_+^d).$$

Since A is in general position $\alpha_1^r \neq 0$ for $r = 1, 2, \ldots, d$ so we may conclude that: (i) φ_- depends (non-trivally) on S_1; that is, on one summand only; and (ii) φ_+ depends (non-trivally) on S_r for $r = 1, 2, \ldots, d$; that is, on d summands. Hence $\varphi = \varphi_- + \varphi_+$ depends non-trivially on $\max\{1, d\}$ summands among the set $\{\mathsf{S}_r : r = 1, 2, \ldots, d\}$.

(2) Assume that exactly two of the coefficients are not zero, say $a_1 \neq 0, a_2 \neq 0$ (without loss of generality). Then

$$\begin{aligned} \varphi_- &= a_1 e_-^1 + a_2 e_-^2 \\ \varphi_+ &= \textstyle\sum_{r=1}^{2} a_r(\alpha_r^1 e_+^1 + \alpha_r^2 e_+^2 + \cdots \alpha_r^d e_+^d). \end{aligned}$$

In this case φ_- depends on two summands S_1 and S_2. Consider the dependence of φ_+ on the summands $\{\mathsf{S}_r : r = 1, 2, \ldots, d\}$ and re-write φ_+ in terms of the specified basis

$$\varphi_+ = \sum_{r=1}^{d} (a_1 \alpha_1^r + a_2 \alpha_2^r) e_+^r.$$

In this last sum suppose two of the basis terms do not appear; that is, suppose the coefficients of e_+^j and e_+^k both vanish, for some $1 \leq j < k \leq d$,

$$\begin{aligned} a_1 \alpha_1^j + a_2 \alpha_2^j &= 0 \\ a_1 \alpha_1^k + a_2 \alpha_2^k &= 0. \end{aligned}$$

But, since the matrix A is in general position,

$$\det \begin{bmatrix} \alpha_1^j & \alpha_2^j \\ \alpha_1^k & \alpha_2^k \end{bmatrix} \neq 0$$

and this leads to the contradiction $a_1 = a_2 = 0$. Therefore, in this case 2, φ_+ depends (non-trivially) on (at least) $d - 1$ summands.

Hence $\varphi = \varphi_- + \varphi_+$ depends (non-trivially) on (at least) $\max\{2, d-1\}$ summands from the set $\{\mathsf{S}_r : r = 1, 2, \ldots, d\}$.

Continue these cases by assuming that three or more of the coefficients $\{a_r : r = 1, 2, \ldots, d\}$ are non-zero.

Then in the (last) case d assume that $a_r \neq 0$ for all $r = 1, 2, \ldots, d$; this assumption gives

$$\begin{aligned} \varphi_- &= \textstyle\sum_{r=1}^{d} a_r e_-^r \\ \varphi_+ &= \textstyle\sum_{r=1}^{d} a_r(\alpha_r^1 e_+^1 + \cdots + \alpha_r^d e_+^d). \end{aligned}$$

Here we note that

φ_- depends on d summands

φ_+ depends on (at least) 1 summand, since $\det[\alpha_j^i] \neq 0$.

Hence $\varphi = \varphi_- + \varphi_+$ depends (non-trivially) on $\max\{d, 1\}$ summands from the set $\{\mathsf{S}_r : r = 1, 2, \ldots, d\}$.

Finally then, examine the list of numbers arising in all these d possible cases, namely $\max\{s, d - (s-1)\}$ for $s = 1, 2, \ldots, d$; the smallest among this list is just $[d/2] + 1$.

Thus we have proved that each non-zero vector $\varphi \in \mathsf{L}_d$ must have non-zero components in (at least) $[d/2] + 1$ of the summands from the set $\{\mathsf{S}_r : r = 1, 2, \ldots, d\}$. Consequently L_d cannot be $[d/2]$-separated.

This completes the construction for Example 7.2.

Remark 7.5. Note that in the discussion of the complete Lagrangian L_d in the complex symplectic space

$$\mathsf{S} = \mathsf{S}_1 \oplus \mathsf{S}_2 \oplus \cdots \oplus \mathsf{S}_d,$$

in Example 7.2, we have demonstrated a much stronger result than asserted in the conclusion of the construction. The requirement on L_d is that each basis should contain at least one vector that couples more than $[d/2]$ summands. We found that every non-zero in L_d couples more than $[d/2]$ summands.

Let us then relate Example 7.2 to the multi-interval system $\{I_r, M_r, w_r : r = 1, 2, \ldots, d\}$ discussed in Remark 7.4 above; there each single-system $\{I_r, M_r, w_r\}$ is regular in L_r^2. Then the complete Lagrangian L_d determines a self-adjoint operator $\mathbf{T}$ on $\mathbf{D}(\mathbf{T}) \subset \mathbf{H} = \sum_{r=1}^{d} \oplus L_r^2$ such that each one of the boundary condition vectors (in any basis for L_d) must involve more than $[d/2]$ of the intervals $\{I_r : r = 1, 2, \ldots, d\}$; that is each boundary condition for $\mathbf{T}$ must couple more than $[d/2]$ of these intervals.

If we bracket or group $\mathsf{S}_1 \oplus \mathsf{S}_2$ together, to play the rôle of S_- and S_+ at the endpoints of I_1, *etc.*, then we obtain corresponding results on the coupling of self-adjoint boundary conditions for regular Sturm-Liouville boundary value problems; compare with the results in [**14**] when $d = 2$.

8. Examples of complete Lagrangians

In Sections 5 and 6 we detailed the process by which a general multi-interval system

$$\{I_r, M_r, w_r : r \in \Omega\} \tag{8.1}$$

generates minimal and maximal operators $\mathbf{T}_0$ on $\mathbf{D}(\mathbf{T}_0)$ and $\mathbf{T}_1$ on $\mathbf{D}(\mathbf{T}_1)$, respectively, in the system Hilbert space $\mathbf{H} = \sum_{r \in \Omega} \oplus L_r^2$; and we established a program of search for self-adjoint extensions $\mathbf{T}$ on $\mathbf{D}(\mathbf{T}) \subset \mathbf{H}$ of the closed symmetric operator $\mathbf{T}_0$. According to our version of the GKN-Theorem 6.2 such self-adjoint extensions $\mathbf{T}$ of $\mathbf{T}_0$ can be described by the complete Lagrangian subspaces of the boundary complex symplectic space

$$\mathsf{S} = \mathbf{D}(\mathbf{T}_1)/\mathbf{D}(\mathbf{T}_0) \approx \mathbf{N}^- \oplus \mathbf{N}^+ \tag{8.2}$$

which is isomorphic to the direct sum of the of the deficiency spaces $\mathbf{N}^\pm$ as in Theorem 6.1. In particular, there exists a complete Lagrangian $\mathsf{L} \subset \mathsf{S}$ if and only if the deficiency indices

$$\mathbf{d}^- = \dim(\mathbf{N}^-) \quad \text{and} \quad \mathbf{d}^+ = \dim(\mathbf{N}^+)$$

are equal, that is, $\mathbf{d} = \mathbf{d}^\pm$ or equivalently, the excess $Ex(\mathsf{S}) = 0$.

In the case when $\dim(\mathsf{S}) < \aleph_0$ the discovery and classification of complete Lagrangians $\mathsf{L} \subset \mathsf{S}$ are effectively treated by matrix calculations within the framework of symplectic algebra, see [**9**]. Further analyses of finite multi-interval systems, with several illuminating examples, are presented in Section 7 above.

However, in the general case where $\dim(\mathsf{S}) \geq \aleph_0$ the existential version of Theorem 6.4 does not offer any explicit construction of the desired complete Lagrangians $\mathsf{L} \subset \mathsf{S}$ and the corresponding self-adjoint operators $\mathbf{T}$ on $\mathbf{D}(\mathbf{T}) \subset \mathbf{H}$.

In this section we shall present explicit descriptions and constructions for infinite dimensional complete Lagrangians L, in infinite dimensional symplectic spaces S, both for abstract algebraic symplectic spaces, and also for boundary complex symplectic spaces of multi-interval systems (8.1), under the necessary condition that the excess $Ex(\mathsf{S}) = 0$.

The constructions for $\mathsf{L} \subset \mathsf{S}$ are done with reference to the corresponding single-interval systems $\{I_r, M_r, w_r\}$, for each fixed $r \in \Omega$, where the corresponding minimal and maximal operators, $T_{0,r}$ on $D(T_{0,r})$ and $T_{1,r}$ on $D(T_{1,r})$ in the Hilbert space $L_r^2 \equiv L^2(I_r; w_r)$ determine the boundary complex symplectic space

$$\mathsf{S}_r = D(T_{1,r})/D(T_{0,r}) \approx N_r^- \oplus N_r^+,$$

with the corresponding deficiency indices $d_r^\pm = \dim(N_r^\pm)$ (so $0 \leq d_r^\pm < \infty$, see Section 2).

It is clear that

$$\mathbf{d}^- = \sum_{r \in \Omega} d_r^- \quad \text{and} \quad \mathbf{d}^+ = \sum_{r \in \Omega} d_r^+ \tag{8.3}$$

can be equal, so $Ex(\mathsf{S}) = 0$, even though every $Ex(\mathsf{S}_r) = d_r^- - d_r^+ \neq 0$ (perhaps all $Ex(\mathsf{S}_r) > 0$, as in Example 6.1 above). Yet it is reasonable to conjecture that whenever $Ex(\mathsf{S}_r) = 0$ for all $r \in \Omega$, the choice of any set of complete Lagrangians $\mathsf{L}_r \subset \mathsf{S}_r$ leads to a complete Lagrangian $\mathsf{L} = \sum_{r \in \Omega} \oplus \mathsf{L}_r$ in the complex symplectic space S.

We shall clarify and confirm this conjecture, based on the Hilbert space construct for

$$\mathsf{S} = \sum_{r \in \Omega} \oplus \mathsf{S}_r, \tag{8.4}$$

and moreover under a relaxation of the demand that $Ex(\mathsf{S}_r) = 0$, for all $r \in \Omega$, to a weaker hypothesis on subfamilies or subcollections of the set of summand spaces $\{\mathsf{S}_r : r \in \Omega\}$, based on the partitions of the index set Ω.

After we obtain quite general results based on the method of partitions, see Definition 8.1 below, we then present several examples which are applications of these constructive procedures. Later we present other kinds of examples where such methods do not apply but, nevertheless, explicit constructions for complete Lagrangians of dimension $\aleph_0$, see [**13**] and [**15**], and some of higher cardinality, can be developed.

DEFINITION 8.1. *A partition* $\{\Omega_\rho : \rho \in \Pi\}$ *of the set* Ω *consists of a collection of non-empty subsets* $\Omega_\rho \subset \Omega$ *(indexed by a set* Π*) which are*

(i) pairwise disjoint, $\Omega_\rho \cap \Omega_\sigma = \emptyset$ *for* $\rho \neq \sigma \in \Pi$

(ii) exhaustive, $\Omega = \bigcup_{\rho \in \Pi} \Omega_\rho$.

Now let $\{I_r, M_r, w_r : r \in \Omega\}$ be a multi-interval system (8.1), and let $\{\Omega_\rho : \rho \in \Pi\}$ be a partition of the index set Ω. Then for each fixed $\rho \in \Pi$ define the subsystem

$$\{I_r, M_r, w_r : r \in \Omega_\rho\}, \tag{8.5}$$

and the corresponding minimal and maximal operators

$$\mathbf{T}_{0,\rho} \text{ on } \mathbf{D}(\mathbf{T}_{0,\rho}) \quad \text{and} \quad \mathbf{T}_{1,\rho} \text{ on } \mathbf{D}(\mathbf{T}_{1,\rho}), \tag{8.6}$$

respectively on the system Hilbert space $\mathbf{H}_\rho = \sum_{r\,\in\,\Omega_\rho} \oplus L^2_r$, as in Section 6 above. Also consider the boundary complex symplectic space (compare Theorem 6.1), for each $\rho \in \Pi$,

$$\mathsf{S}_\rho = \mathbf{D}(\mathbf{T}_{1,\rho})/\mathbf{D}(\mathbf{T}_{0,\rho}) \approx \mathbf{N}^-_\rho \oplus \mathbf{N}^+_\rho. \tag{8.7}$$

Remark 8.1. We note how the operators and spaces relevant for the multi-interval system (8.1) can be restricted to a subsystem (8.5) to obtain their analogues. For instance, for each fixed $\rho \in \Pi$, with respect to the partition $\{\Omega_\rho : \rho \in \Pi\}$ for Ω, the system Hilbert space $\mathbf{H}_\rho$ can be considered as a closed subspace of the Hilbert space $\mathbf{H}$, namely

$$\mathbf{H}_\rho = \{\mathbf{f} \in \mathbf{H} : f_r = 0 \text{ for } r \notin \Omega_\rho\}, \tag{8.8}$$

and the norm $\|\cdot\|_{\mathbf{H}_\rho}$ coincides with $\|\cdot\|_{\mathbf{H}}$, when restricted to $\mathbf{H}_\rho$. Moreover $\mathbf{H}_\rho$ and $\mathbf{H}_\sigma$ are orthogonal subspaces of $\mathbf{H}$ for $\rho \neq \sigma \in \Pi$. Then it follows immediately that $\mathbf{T}_{0,\rho}$ and $\mathbf{T}_{1,\rho}$ (adjoints of each other as operators on $\mathbf{H}_\rho$) are both restrictions of $\mathbf{T}_1$ to their corresponding domains:

$$\mathbf{D}(\mathbf{T}_{0,\rho}) = \mathbf{D}(\mathbf{T}_0) \cap \mathbf{H}_\rho, \qquad \mathbf{D}(\mathbf{T}_{1,\rho}) = \mathbf{D}(\mathbf{T}_1) \cap \mathbf{H}_\rho. \tag{8.9}$$

Also the deficiency spaces are

$$\mathbf{N}^\pm_\rho = \mathbf{N}^\pm \cap \mathbf{H}_\rho \text{ with deficiency indices } \mathbf{d}^\pm_\rho = \dim(\mathbf{N}^\pm_\rho), \tag{8.10}$$

for each $\rho \in \Pi$, as before.

Further we identify (just as for the full index set Ω in (8.1))

$$\mathbf{N}^\pm_\rho = \sum_{r\,\in\,\Omega_\rho} \oplus N^\pm_r, \qquad \mathsf{S}_\rho = \sum_{r\,\in\,\Omega_\rho} \oplus \mathsf{S}_r, \tag{8.11}$$

as complex symplectic spaces and also as Hilbert spaces, as in Theorem 6.3. Again we have the (cardinal number) sum - compare Corollary 6.2,

$$\mathbf{d}^\pm_\rho = \sum_{r\,\in\,\Omega_\rho} d^\pm_r. \tag{8.12}$$

The same considerations show that for the complex symplectic space S_ρ we have the equalities for the symplectic product

$$[\mathsf{f} : \mathsf{g}]_{\mathsf{S}_\rho} = \sum_{r\,\in\,\Omega_\rho} [f_r : g_r]_{D_r} \quad \text{(or equivalently } \sum_{r\,\in\,\Omega_\rho} [\mathsf{f}_r : \mathsf{g}_r]_{\mathsf{S}_r}) \tag{8.13}$$

and for the Hilbert space scalar product

$$\langle \mathsf{f}, \mathsf{g} \rangle_{\mathsf{S}_\rho} = \sum_{r\,\in\,\Omega_\rho} \langle f_r : g_r \rangle_{D_r} \quad \text{(or equivalently } \sum_{r\,\in\,\Omega_\rho} \langle \mathsf{f}_r, \mathsf{g}_r \rangle_{\mathsf{S}_r}), \tag{8.14}$$

for vectors in S_ρ

$$\mathsf{f} = \{\mathsf{f}_r \in \mathsf{S}_r : r \in \Omega_\rho\}, \quad \mathsf{g} = \{\mathsf{g}_r \in \mathsf{S}_r : r \in \Omega_\rho\},$$

where $\mathsf{f}_r, \mathsf{g}_r \in \mathsf{S}_r$ correspond to functions $f_r, g_r \in N^-_r \oplus N^+_r$, as usual.

These calculations are valid because all these sums are absolutely convergent, and involve only countably many non-zero terms. Furthermore all the norms indicated in this Remark 8.1 are equivalent to the $\mathbf{H}$-norm restricted to $\mathbf{H}_\rho$ (recall that convergence in $\mathsf{S} \approx \mathbf{N}^- \oplus \mathbf{N}^+$ in the $\mathbf{D}$-metric is equivalent to simultaneous convergence in $\mathbf{N}^\pm$ in the $\mathbf{H}$-metric, see (6.15)).

This concludes Remark 8.1, concerning the space S_ρ for each $\rho \in \Pi$.

THEOREM 8.1. *Let $\{I_r, M_r, w_r : r \in \Omega\}$ be a multi-interval system, as in Definition 2.1, with the boundary complex symplectic space*

$$\mathsf{S} = \mathbf{D}(\mathbf{T}_1)/\mathbf{D}(\mathbf{T}_0) \approx \mathbf{N}^- \oplus \mathbf{N}^+,$$

as in Theorem 6.1.

Let $\{\Omega_\rho : \rho \in \Pi\}$ be a partition of Ω, see Definition 8.1, and consider the corresponding subsystems $\{I_r, M_r, w_r : r \in \Omega_\rho\}$, for each fixed $\rho \in \Pi$, with the boundary complex symplectic space

$$\mathsf{S}_\rho = \mathbf{D}(\mathbf{T}_{1,\rho})/\mathbf{D}(\mathbf{T}_{0,\rho}) \approx \mathbf{N}_\rho^- \oplus \mathbf{N}_\rho^+, \tag{8.15}$$

as defined above in (8.7).

Then

$$\mathbf{N}^\pm = \sum_{\rho \in \Pi} \oplus \mathbf{N}_\rho^\pm \quad \textit{and} \quad \mathsf{S} = \sum_{\rho \in \Pi} \oplus \mathsf{S}_\rho \tag{8.16}$$

and the corresponding symplectic and scalar products satisfy

$$[\mathsf{f} : \mathsf{g}]_\mathsf{S} = \sum_{\rho \in \Pi} [\mathsf{f}_\rho : \mathsf{g}_\rho]_{\mathsf{S}_\rho}, \quad \langle \mathsf{f}, \mathsf{g} \rangle_\mathsf{S} = \sum_{\rho \in \Pi} \langle \mathsf{f}_\rho, \mathsf{g}_\rho \rangle_{\mathsf{S}_\rho}, \tag{8.17}$$

for all $\mathsf{f} = \{\mathsf{f}_\rho \in \mathsf{S}_\rho : \rho \in \Pi\}, \mathsf{g} = \{\mathsf{g}_\rho \in \mathsf{S}_\rho : \rho \in \Pi\}$ in S. Also the deficiency indices satisfy

$$\mathbf{d}^\pm = \sum_{\rho \in \Pi} \mathbf{d}_\rho^\pm. \tag{8.18}$$

Hence if $Ex(\mathsf{S}_\rho) = 0$ for all $\rho \in \Pi$, then $Ex(\mathsf{S}) = 0$.

PROOF. Convergence in the space $\sum_{\rho \in \Pi} \oplus \mathbf{N}_\rho^+$ refers to the topology of the $\mathbf{H}_\rho$-norm, which coincides with the $\mathbf{H}$-norm on the Hilbert space $\mathbf{H}_\rho$. Hence $\sum_{\rho \in \Pi} \oplus \mathbf{N}_\rho^+ \subseteq \mathbf{N}^+$.

Now note that $\mathbf{N}_\rho^+$ and $\mathbf{N}_\sigma^+$ are orthogonal for $\rho \neq \sigma \in \Pi$. Furthermore, if a vector $\mathbf{u}_+ \in \mathbf{N}^+$ is $\mathbf{H}$-orthogonal to $\mathbf{N}_\rho^+ = \sum_{r \in \Omega_\rho} \oplus \mathbf{N}_r^+$, then $\mathbf{u}_+$ is orthogonal to N_r^+ (in the Hilbert space $\mathbf{H}$) for each $r \in \Omega_\rho$. Hence, in this case, $\mathbf{u}_+$ being $\mathbf{H}$-orthogonal to each $\mathbf{N}_\rho^+$ for $\rho \in \Pi$, implies that $\mathbf{u}_+$ is orthogonal to each $\mathbf{N}_r^+$ for $r \in \Omega$, and thus $\mathbf{u}_+ = 0$. Therefore we conclude that $\mathbf{N}^+ = \sum_{\rho \in \Pi} \mathbf{N}_\rho^+$. A similar result holds for $\mathbf{N}^-$, and consequently for $\mathsf{S} \approx \mathbf{N}^- \oplus \mathbf{N}^+$.

The remaining equalities of (8.17) then follow directly from the absolute convergence of these sums. □

THEOREM 8.2. *Let $\{I_r, M_r, w_r : r \in \Omega\}$ be a multi-interval system and $\{\Omega_\rho : \rho \in \Pi\}$ be a partition of Ω, as in Theorem 8.1.*

Assume that each subsystem $\{I_r, M_r, w_r : r \in \Omega_\rho\}$ for fixed $\rho \in \Pi$, has a boundary complex symplectic space

$$\mathsf{S}_\rho = \sum_{r \in \Omega_\rho} \oplus \mathsf{S}_r \quad \textit{with} \quad Ex(\mathsf{S}_\rho) = 0. \tag{8.19}$$

Let $\mathsf{L}_\rho \subset \mathsf{S}_\rho$ *be a complete Lagrangian subspace of* S_ρ*, for each* $\rho \in \Pi$*, and define*

$$\mathsf{L} := \sum_{\rho \in \Pi} \oplus \mathsf{L}_\rho, \tag{8.20}$$

as the direct sum of the Hilbert spaces L_ρ *of* S*. Then* L *is a complete Lagrangian subspace of* S*.*

Further we can characterize L *either by (i) or (ii) as follows:*

(i) The closure (in the Hilbert space S*) of the Lagrangian submanifold* $\mathsf{L}_0 \subset \mathsf{S}$*, where*

$$\mathsf{L}_0 := \operatorname{span}\{\mathsf{L}_\rho : \rho \in \Pi\};$$

that is, L_0 *consists of all finite sums of vectors of* $\bigcup_{\rho \in \Pi} \mathsf{L}_\rho$*, and then*

$$\mathsf{L} = \overline{\operatorname{span}}\{\mathsf{L}_\rho : \rho \in \Pi\}.$$

(ii) L *is the unique Hilbert subspace of* S *generated by all* L_ρ *for* $\rho \in \Pi$*; that is, considering*

$$\mathsf{L}_\rho \subset \mathsf{S}_\rho = \sum_{r \in \Omega_\rho} \oplus \mathsf{S}_r \subset \mathsf{S},$$

define L *as the intersection of all Hilbert subspaces of* S*, which each contains all* L_ρ *for* $\rho \in \Pi$*.*

Moreover, an orthonormal basis of L *can be can be constructed as a union of orthonormal bases of* L_ρ*, choosing one such basis for each* L_ρ*,* $\rho \in \Pi$*.*

PROOF. Since each S_ρ, for $\rho \in \Pi$, has $Ex(\mathsf{S}_\rho) = 0$, then $Ex(\mathsf{S}) = 0$ and $\mathbf{d} = \mathbf{d}^\pm$ by Theorem 8.1.

Let L_ρ be a complete Lagrangian in the complex symplectic space S_ρ, for each $\rho \in \Pi$. Then L_ρ is a closed Hilbert subspace of S_ρ and hence a closed subspace of S, see (8.16). Thus we can construct the appropriate direct sum Hilbert space

$$\mathsf{L} = \sum_{\rho \in \Pi} \oplus \mathsf{L}_\rho,$$

which is a Lagrangian submanifold of S.

We shall show that L is a closed Hilbert subspace of $\mathsf{S} \approx \mathbf{N}^- \oplus \mathbf{N}^+$, and, in fact, a complete Lagrangian subspace of S.

Assume then that a vector $\mathsf{f} = \{\mathbf{f}_- + \mathbf{f}_+ + \mathbf{D}(\mathbf{T}_0)\} \in \mathsf{S}$ satisfies the condition $[\mathsf{f} : \mathsf{L}]_\mathsf{S} = 0$, for some choice of $\mathbf{f}_\pm \in \mathbf{N}^\pm \subset \mathbf{D}(\mathbf{T}_1)$.

Take a vector $\mathsf{v} = \{\mathsf{v}_\rho \in \mathsf{S}_\rho : \rho \in \Pi\}$ and then $[\mathsf{f} : \mathsf{v}]_\mathsf{S} = \sum_{\rho \in \Pi} [\mathsf{f}_\rho : \mathsf{v}_\rho]_{\mathsf{S}_\rho}$. Now select $\mathsf{v} \in \mathsf{S}$ so that all components are zero except for v_{ρ_1}, which can be taken arbitrarily within S_{ρ_1}; in particular take $\mathsf{v}_{\rho_1} \in \mathsf{L}_{\rho_1}$. Then since $[\mathsf{f} : \mathsf{L}]_\mathsf{S} = 0$, and $\mathsf{v} \in \mathsf{L}_{\rho_1} \subset \mathsf{L}$ it follows that $[\mathsf{f} : \mathsf{v}]_\mathsf{S} = [\mathsf{f}_{\rho_1} : \mathsf{v}_{\rho_1}]_{\mathsf{S}_{\rho_1}} = 0$. Therefore $[\mathsf{f}_{\rho_1} : \mathsf{L}_{\rho_1}]_{\mathsf{S}_{\rho_1}} = 0$ and $\mathsf{f}_{\rho_1} \in \mathsf{L}_{\rho_1}$, since L_{ρ_1} is a complete Lagrangian subspace of S_{ρ_1}.

Hence we conclude that $\mathsf{f} = \{\mathsf{f}_\rho \in \mathsf{L}_\rho : \rho \in \Pi\} \in \mathsf{S}$ must have each component $\mathsf{f}_\rho \in \mathsf{L}_\rho$ and thus $\mathsf{f} \in \mathsf{L} = \sum_{\rho \in \Pi} \oplus \mathsf{L}_\rho$. Therefore L is a complete Lagrangian subspace of S, and so L is a closed Hilbert subspace of S.

The characterization (*i*) $\mathsf{L} = \overline{\operatorname{span}}\{\mathsf{L}_\rho : \rho \in \Pi\}$ is straightforward. Also the characterization (*ii*) for L as the intersection of all Hilbert subspaces of S, which each contains every subspace L_ρ (considered as a closed subspace $\mathsf{L}_\rho \subset \mathsf{S}_\rho \subset \mathsf{S}$), is then an immediate consequence. □

COROLLARY 8.1. *Let $\{I_r, m_r, w_r : r \in \Omega\}$ be a multi-interval system, and let $\{\Omega_\rho : \rho \in \Pi\}$ be a partition of Ω, so as to define the boundary complex symplectic spaces S_ρ,*

$$\mathsf{S}_\rho = \sum_{r \in \Omega_\rho} \oplus \mathsf{S}_r \quad \textit{and further} \quad \mathsf{S} = \sum_{\rho \in \Pi} \oplus \mathsf{S}_\rho$$

as in Theorem 8.2. Also assume that $Ex(\mathsf{S}_\rho) = 0$, for $\rho \in \Pi$, so $Ex(\mathsf{S}) = 0$; and specify a complete Lagrangian subspace $\mathsf{L}_\rho \subset \mathsf{S}_\rho$ for each $\rho \in \Pi$.

Then, by Theorem 8.2, $\mathsf{L} = \sum_{\rho \in \Pi} \oplus \mathsf{L}_\rho$ is a complete Lagrangian subspace of S. Now denote the corresponding self-adjoint operators (as in the GKN-Theorem 6.2) by

$$\mathbf{T}_\rho \textit{ on } \mathbf{D}(\mathbf{T}_\rho) \subset \mathbf{H}_\rho \quad \textit{and} \quad \mathbf{T} \textit{ on } \mathbf{D}(\mathbf{T}) \subset \mathbf{H}.$$

Then

$$\mathbf{D}(\mathbf{T})/\mathbf{D}(\mathbf{T}_0) = \sum_{\rho \in \Pi} \oplus \left(\mathbf{D}(\mathbf{T}_\rho)/\mathbf{D}(\mathbf{T}_{0,\rho})\right) \tag{8.21}$$

so

$$\mathbf{D}(\mathbf{T}) = \left(\sum_{\rho \in \Pi} \oplus \left(\mathbf{D}(\mathbf{T}_\rho)/\mathbf{D}(\mathbf{T}_{0,\rho})\right)\right) \oplus \mathbf{D}(\mathbf{T}_0), \tag{8.22}$$

where we identify

$$\mathbf{D}(\mathbf{T}_\rho)/\mathbf{D}(\mathbf{T}_{0,\rho}) = \mathsf{L}_\rho \subset \mathsf{S}_\rho \approx \mathbf{N}_\rho^- \oplus \mathbf{N}_\rho^+ \tag{8.23}$$

and thus

$$\mathsf{L} = \sum_{\rho \in \Pi} \oplus \left(\mathbf{D}(\mathbf{T}_\rho)/\mathbf{D}(\mathbf{T}_{0,\rho})\right) \subset \sum_{\rho \in \Pi} \oplus \left(\mathbf{N}_\rho^- \oplus \mathbf{N}_\rho^+\right) = \mathbf{N}^- \oplus \mathbf{N}^+.$$

PROOF. The corollary follows directly from Theorems 8.1 and 8.2. □

We now use the method of partitions, Definition 8.1 and the resulting Theorems 8.1 and 8.2 above, to construct an example of an infinite dimensional complete Lagrangian L_∞, in a complex symplectic space $\mathsf{S} = \sum_{r=1}^{\infty} \oplus \mathsf{S}_r$, such that L_∞ is not k-separated for any positive integer $k \geq 1$, that is, L_∞ is *not finitely separated* as explained below in Definition 8.2. We follow the notations and techniques introduced in the constructions for Example 7.2 above, where we constructed a complete Lagrangian L_d in the finite dimensional complex symplectic space $\mathsf{S} = \sum_{r=1}^{d} \oplus \mathsf{S}_r$, which is not $[d/2]$-separated. However for the infinite dimensional case we must first resolve and clarify certain problems of topology and corresponding questions concerning convergence.

Let Ω be a non-empty index set, of arbitrary cardinality, and for each $r \in \Omega$ we specify a complex symplectic space S_r which is also a Hilbert space, say as described in Example 3.1, or as occurs for multi-interval systems $\{I_r, M_r, w_r : r \in \Omega\}$ in Section 6 above. Then consider the Hilbert space direct sum

$$\mathsf{S} = \sum_{r \in \Omega} \oplus \mathsf{S}_r, \tag{8.24}$$

which is also a complex symplectic space, as well as a Hilbert space, with the symplectic and scalar products inter-related as usual; see Theorem 6.3. In this current discussion we deal only with the abstract algebraic and topological properties of S, and leave until later the description of the analytical theory for the corresponding multi-interval systems; compare the Remarks 7.4 and 7.5 above.

DEFINITION 8.2. *Consider a complex symplectic space*

$$\mathsf{S} = \sum_{r \in \Omega} \oplus \mathsf{S}_r$$

which is also a Hilbert space direct sum of the given complex symplectic summands (also Hilbert spaces) S_r *for* $r \in \Omega$, *as in* (8.24).

Then a complete Lagrangian subspace $\mathsf{L} \subset \mathsf{S}$ *is called* k*-separated, for a given integer* $k \geq 1$, *in case*

$$\mathsf{L} = \overline{\text{span}} \left\{ \mathsf{L} \cap \left(\bigcup_{\substack{distinct \\ \{r_1, r_2, \ldots, r_k\}}} (\mathsf{S}_{r_1} \oplus \mathsf{S}_{r_2} \oplus \cdots \oplus \mathsf{S}_{r_k}) \right) \right\}, \tag{8.25}$$

where the union is taken over all possible k*-tuples* $\{r_1, r_2, \ldots, r_k\}$ *of distinct indices in* Ω, *and the closure of the span is relative to the topology of the Hilbert subspace* L *in* S.

If there exists some $k \geq 1$ *for which* L *is* k*-separated, then* L *is said to be* ***finitely separated*** *in* S. *On the other hand if, for each integer* $k \geq 1$, L *is not* k*-separated then* L *is said to be* ***not finitely separated*** *in* S.

Remark 8.2. We rephrase Definition 8.2 above:

L is k-separated just in case there exists a subset of vectors $\{\varphi_\alpha\}$ in L, with each φ_α in some k-tuple direct summand $\mathsf{S}_{r_1} \oplus \mathsf{S}_{r_2} \oplus \cdots \oplus \mathsf{S}_{r_k}$ (possibly different vectors in different k-summands), so that each vector $\mathsf{v} \in \mathsf{L}$ can be arbitrarily closely approximated by finite linear combinations of the given subset $\{\varphi_\alpha\}$.

In case $\dim(\mathsf{S}) < \infty$, we can assert that L is k-separated if and only if there exists a basis for L, such that each of these basis vectors lies in some k-tuple of direct sums $\mathsf{S}_{r_1} \oplus \mathsf{S}_{r_2} \oplus \cdots \oplus \mathsf{S}_{r_k}$. Hence Definition 8.2 is entirely equivalent to the conditions of Definition 7.1 for finite dimensional complex symplectic spaces.

Note that in either case the vectors $\{\varphi_\alpha\}$ need not contain an orthonormal basis for the Hilbert space L.

Example 8.1. Consider the complex symplectic space

$$\mathsf{S} = \sum_{r=1}^{\infty} \oplus \mathsf{S}_r, \tag{8.26}$$

with each symplectic space S_r, for $r \in \Omega = \mathbb{N}$, defined by the invariants

$$\dim(\mathsf{S}_r) = 2, \quad Ex(\mathsf{S}_r) = 0 \quad \text{for } r \in \mathbb{N}. \tag{8.27}$$

Moreover, for each $r \in \mathbb{N}$ select a basis $\{e_-^r, e_+^r\}$ for S_r with

$$[e_-^r : e_-^r] = -i, \quad [e_+^r : e_+^r] = i, \quad [e_-^r : e_+^r] = 0; \tag{8.28}$$

and further specify the Hilbert space norm in S_r by asserting that $\{e_-^r, e_+^r\}$ is an orthonormal basis.

In this way each S_r is determined, up to symplectic and unitary isomorphism, and hence the complex symplectic S, with Hilbert space norm, is also uniquely determined according to Theorem 6.3 above. Clearly

$$\dim(\mathsf{S}) = \aleph_0 \quad \text{and} \quad Ex(\mathsf{S}) = 0. \tag{8.29}$$

We construct a complete Lagrangian L_∞ which is not k-separated in S, and this for every prescribed integer $k \geq 1$. Hence L_∞ will be an infinite dimensional complete Lagrangian in S, and L_∞ is not finitely separated in S. The method of construction for L_∞ is to collect together long strings of summands from $\{\mathsf{S}_r : r \in \Omega\}$ into symplectic spaces $\tilde{\mathsf{S}}_\rho$ (corresponding to a partition $\{\Omega_\rho : \rho \in \Pi\}$ of $\Omega = \mathbb{N}$) and then apply the techniques developed in Example 7.2 to the finite dimensional spaces $\tilde{\mathsf{S}}_\rho$.

For this purpose partition $\Omega = \mathbb{N}$ into subsets Ω_ρ for $\rho \in \Pi = \mathbb{N}$ (see Definition 8.1): namely,

$$\Omega_1 = \{1\}, \quad \Omega_2 = \{2,3\}, \quad \Omega_3 = \{4,5,6,7\},$$

and generally for each $\rho \in \Pi = \mathbb{N}$

$$\Omega_\rho = \{2^{\rho-1}, \dots, 2^\rho - 1\}. \tag{8.30}$$

Then denote the corresponding symplectic spaces $\tilde{\mathsf{S}}_\rho$ for $\rho \in \Pi = \mathbb{N}$ accordingly,

$$\tilde{\mathsf{S}}_1 = \mathsf{S}_1, \quad \tilde{\mathsf{S}}_2 = \mathsf{S}_2 \oplus \mathsf{S}_3, \quad \tilde{\mathsf{S}}_3 = \mathsf{S}_4 \oplus \mathsf{S}_5 \oplus \mathsf{S}_6 \oplus \mathsf{S}_7$$

and generally for each $\rho \in \mathbb{N}$

$$\tilde{\mathsf{S}}_\rho = \mathsf{S}_{2^{\rho-1}} \oplus \cdots \oplus \mathsf{S}_{2^\rho - 1}, \tag{8.31}$$

so $\tilde{\mathsf{S}}_\rho$ is the direct sum of $2^{\rho-1}$ summands of $\{\mathsf{S}_r : r \in \mathbb{N}\}$, and $\tilde{\mathsf{S}}_\rho$ is a Hilbert subspace of S. In particular $\tilde{\mathsf{S}}_\rho$ has a basis of 2^ρ vectors, namely

$$\left\{ e_-^{2^{\rho-1}}, e_+^{2^{\rho-1}}, \dots, e_-^{2^\rho-1}, e_+^{2^\rho-1} \right\}.$$

Next we decompose each $\tilde{\mathsf{S}}_\rho$ into the direct sum of Hilbert spaces,

$$\tilde{\mathsf{S}}_\rho = \tilde{H}_-^\rho \oplus \tilde{H}_+^\rho \tag{8.32}$$

where $\tilde{H}_\pm^\rho$ have the corresponding orthonormal bases specifed by

$$\left\{ e_-^{2^{\rho-1}}, \dots, e_-^{2^\rho-1} \right\} \quad \text{and} \quad \left\{ e_+^{2^{\rho-1}}, \dots, e_+^{2^\rho-1} \right\}. \tag{8.33}$$

Following the program of Example 7.2 above, we define a unitary map U^ρ of $\tilde{H}_-^\rho$ onto $\tilde{H}_+^\rho$, for each $\rho \in \mathbb{N}$, by its action on the orthonormal basis $\left\{ e_-^r : r = 2^{\rho-1}, \dots, 2^\rho - 1 \right\}$,

$$U^\rho : e_-^r \to e_+^r \quad \text{for} \quad r = 2^{\rho-1}, \dots, 2^\rho - 1. \tag{8.34}$$

Now compose the map U^ρ with a "general position rotation map"

$$\tilde{A}^\rho : \tilde{H}_+^\rho \to \tilde{H}_+^\rho, \quad e_+^r \to \tilde{e}_+^r \quad \text{for} \quad r = 2^{\rho-1}, \dots, 2^\rho - 1, \tag{8.35}$$

in accord with the concepts of Definition 7.2 above. Namely, select a rotation matrix $\tilde{A}^\rho \in SO(2^{\rho-1}, \mathbb{R})$ which is in general position in this special orthogonal group (see Definition 7.2), and use this rotation matrix to define the rotation map (also denoted by $\tilde{A}^\rho$), after coordinatizing $\tilde{H}_+^\rho$ so that $e_+^{2^{\rho-1}} = (1,0,0,\dots,0)^t, \cdots, e_+^{2^\rho-1} = (0,0,\dots,0,1)^t$. That is, define

$$\tilde{e}_+^r := \tilde{A}^\rho e_+^r \quad \text{for} \quad r = 2^{\rho-1}, \dots, 2^\rho - 1. \tag{8.36}$$

In this way we define the rotation map $\tilde{A}^\rho$ which is a unitary surjection of $\tilde{H}_+^\rho$ onto itself.

Now define the unitary surjection of $\tilde{H}_-^\rho$ onto $\tilde{H}_+^\rho$:

$$\tilde{U}^\rho : \tilde{H}_-^\rho \to \tilde{H}_+^\rho, \quad e_-^r \to \tilde{e}_+^r \quad \text{for} \quad r \in \Omega_\rho, \tag{8.37}$$

that is,

$$\tilde{U}^\rho = \tilde{A}^\rho U^\rho \quad \text{for each} \quad \rho \in \Pi.$$

As usual we define the Hilbert subspaces of S,

$$H_- := \overline{\text{span}}\{e_-^r : r \in \mathbb{N}\}, \quad H_+ := \overline{\text{span}}\{e_+^r : r \in \mathbb{N}\}; \tag{8.38}$$

thus

$$H_\pm = \sum_{\rho \in \mathbb{N}} \oplus \tilde{H}_\pm^\rho \quad \text{and} \quad \mathsf{S} = H_- \oplus H_+.$$

Then the collection of maps $\tilde{U}^\rho$ for $\rho \in \Pi = \mathbb{N}$ defines a unitary surjection of H_- onto H_+, namely

$$U_\infty : H_- \to H_+, \quad \text{with} \quad U_\infty e_-^r := \tilde{U}^\rho e_-^r = \tilde{e}_+^r, \quad \text{for} \quad r \in \Omega_\rho. \tag{8.39}$$

That is, U_∞ carries the orthonormal basis $\{e_-^r : r \in \mathbb{N}\}$ in H_- onto the orthonormal basis $\{\tilde{e}_+^r : r \in \mathbb{N}\}$ in H_+.

The required complete Lagrangian $\mathsf{L}_\infty \subset \mathsf{S}$ is then defined by

$$L_\infty := \text{graph}(U_\infty), \tag{8.40}$$

see extension of [**9**, Section III.1, Lemma 1, Page 34] to infinite dimensional complex symplectic spaces [**11**]. That is a vector $\mathsf{v} \in \mathsf{S}$ has the unique decomposition $\mathsf{v} = \mathsf{v}_- + \mathsf{v}_+$ with $\mathsf{v}_\pm \in H_\pm$, and $\mathsf{v} \in \mathsf{L}_\infty$ if and only if $\mathsf{v}_+ = U_\infty \mathsf{v}_-$ in S.

We now demonstrate that, for each integer $k \geq 1$, L_∞ is not k-separated, and hence that L_∞ is an infinite dimensional complete Lagrangian which is not finitely separated in S.

Accordingly, fix a positive integer $k \geq 1$, and we then show that there exists a vector $\mathsf{v} \in \mathsf{L}_\infty$ (possibly v depends on k), which cannot be approximated by linear combinations of vectors $\{\varphi_\alpha\}$ in L_∞, each of which is (at most) k-coupled to the subspaces $\{\mathsf{S}_r : r \in \mathbb{N}\}$.

Make the explicit choice $\mathsf{v} = e_-^{2^{\rho-1}} + \tilde{e}_+^{2^{\rho-1}} \in \mathsf{L}_\infty$, with ρ chosen so that $2^{\rho-2} \geq k$, say for definiteness $\rho = [\log_2(k) + 3]$, *i.e.* integral part of $(\log_2(k) + 3)$. We also use the fact that

$$\tilde{\mathsf{S}}_\rho = \mathsf{S}_{2^{\rho-1}} \oplus \cdots \oplus \mathsf{S}_{2^\rho - 1}$$

involves $2^{\rho-1} \geq 2k$ summands.

Suppose there does exist a finite set of vectors $\{\varphi_s : s = 1, 2, \ldots, m\}$ in L_∞, each of which depends non-trivially on (at most) k of the summands $\{\mathsf{S}_r : r \in \mathbb{N}\}$, and such there is a linear combination

$$\sum_{s=1}^m \beta_s \varphi_s \quad \text{with} \quad \beta_s \in \mathbb{C} \quad \text{for} \quad s = 1, 2, \ldots, m$$

that approximates v to within an error less than $1/2$ in the metric of S. In such a case at least one of these vectors, call it φ, must have a non-zero coefficient in the term $e_-^{2^{\rho-1}}$ when φ is expanded along the orthonormal basis $\{e_\pm^r : r \in \mathbb{N}\}$ of S.

Consider φ and decompose it as

$$\varphi = \varphi_- + \varphi_+, \quad \text{with} \quad \varphi_\pm \in H_\pm$$

so that, since $\varphi \in \mathsf{L}_\infty$, $\varphi_+ = U_\infty \varphi_-$. Then we expand each of $\varphi_\pm$ in terms of the basis $\{e^r_\pm : r \in \mathbb{N}\}$ to write

$$\begin{aligned}\varphi_- &= \sum_{s=1}^{2^{\rho-1}} a_s e_-^{2^{\rho-1}+(s-1)} + (\text{terms off } \tilde{\mathsf{S}}_\rho) \\ &:= \tilde{\varphi}^\rho_- + (\text{terms off } \tilde{\mathsf{S}}_\rho)\end{aligned} \tag{8.41}$$

and

$$\begin{aligned}\varphi_+ &= \sum_{s=1}^{2^{\rho-1}} a_s e_+^{2^{\rho-1}+(s-1)} + (\text{terms off } \tilde{\mathsf{S}}_\rho) \\ &:= \tilde{\varphi}^\rho_+ + (\text{terms off } \tilde{\mathsf{S}}_\rho)\end{aligned}$$

for coefficients $\{a_s \in \mathbb{C} : s = 1, 2, \ldots, 2^{\rho-1}\}$ with $a_1 \neq 0$. Thus $\tilde{\varphi}^\rho_- \neq 0$, and

$$\tilde{\varphi}^\rho := \tilde{\varphi}^\rho_- + \tilde{\varphi}^\rho_+ \in \operatorname{graph}\left(\tilde{U}^\rho\right). \tag{8.42}$$

Using the argument of Example 7.2 and the subsequent Remark 7.5, as applied to $\tilde{\varphi}^\rho \in \tilde{\mathsf{S}}_\rho$, we conclude that $\tilde{\varphi}^\rho$ must depend non-trivially on (at least) $1+2^{\rho-1}/2 \geq k+1$ of the summands $\mathsf{S}_{2^{\rho-1}}, \ldots, \mathsf{S}_{2^\rho - 1}$. But this contradicts the assumptions on the vector φ.

Hence we conclude that

$$\mathsf{v} \notin \mathsf{L} = \overline{\operatorname{span}}\left\{ \mathsf{L}_\infty \cap \left(\bigcup_{\substack{\text{distinct}\\ \{r_1, r_2, \ldots, r_k\}}} (\mathsf{S}_{r_1} \oplus \mathsf{S}_{r_2} \oplus \cdots \oplus \mathsf{S}_{r_k}) \right) \right\},$$

and therefore L_∞ is not k-separated in S. Since $k \geq 1$ is an arbitrary integer, we conclude that L_∞ is not finitely separated in $\mathsf{S} = \sum_{r=1}^\infty \oplus \mathsf{S}_r$. Thus the required properties of L_∞ have been verified, and the construction for Example 8.1 is completed.

Remark 8.3. It is also possible to construct L_∞ in Example 8.1 as the direct sum of the finite dimensional Lagrangians

$$\tilde{\mathsf{L}}_\rho := \operatorname{graph}\left(\tilde{U}^\rho\right) \quad \text{in} \quad \tilde{\mathsf{S}}_\rho \quad \text{for} \quad \rho \in \Pi = \mathbb{N}. \tag{8.43}$$

That is,

$$\mathsf{L}_\infty = \sum_{\rho=1}^\infty \oplus \tilde{\mathsf{L}}_\rho, \tag{8.44}$$

with the infinite sum convergent in $\mathsf{S} = \sum_{\rho \in \Pi} \oplus \tilde{\mathsf{S}}_\rho$, as in Theorems 8.1 and 8.2 above.

This last section of the paper is closed with several examples of multi-interval systems $\{I_r, M_r, w_r : r \in \Omega\}$, as in (8.1), significantly some where Ω is an infinite set, *i.e.* $\operatorname{card}(\Omega) \geq \aleph_0$. In particular, we describe various complete Lagrangians L for the corresponding boundary complex symplectic space $\mathsf{S} = \mathbf{D}(\mathbf{T}_1)/\mathbf{D}(\mathbf{T}_0)$, and hence determine self-adjoint extensions $\mathbf{T}$ of the minimal operator $\mathbf{T}_0$, as in the GKN-Theorem 6.2.

In certain of these examples it will be useful to identify each interval I_r, for $r \in \Omega$, with the corresponding interval on a fixed or reference real line $\mathbb{R}$, especially when $\text{card}(\Omega) = \aleph_0$ and the corresponding intervals are piecewise disjoint on $\mathbb{R}$ or intersect only at common endpoints, compare Definition 2.1(a).

Our methods will then refer to the construction of a complete Lagrangian L in the complex symplectic space $\mathsf{S} = \sum_{r \in \Omega} \oplus \mathsf{S}_r$ where $\mathsf{S} \approx \mathbf{N}^- \oplus \mathbf{N}^+$ and $\mathsf{S}_r \approx N_r^- \oplus N_r^+$, as in Theorem 6.3 and its Corollary 6.2, above. Namely, an orthonormal basis for the Hilbert space $\mathbf{N}^- = \sum_{r \in \Omega} \oplus N_r^-$ can be selected as the union of orthonormal bases for all N_r^-, where $r \in \Omega$, and the corresponding result for $\mathbf{N}^+$.

Now there exists a one-to-one correspondence between the set $\{\mathsf{L}\}$ of all complete Lagrangians in S and the set $\{\mathbf{U}\}$ of all unitary maps of $\mathbf{N}^-$ onto $\mathbf{N}^+$, more explicitly

$$\mathsf{L} = \text{graph}(\mathbf{U}), \tag{8.45}$$

as in (8.40) above. This means that for each chosen bijective correspondence between an orthonormal basis for $\mathbf{N}^-$ and an orthonormal basis for $\mathbf{N}^+$, there corresponds such a unitary surjection $\mathbf{U}$ of $\mathbf{N}^-$ onto $\mathbf{N}^+$, and hence a complete Lagrangian $\mathsf{L} \subset \mathsf{S}$ according to (8.45). This construction for L can be carried out successfully even though each pair N_r^- and N_r^+ have different dimensions ($Ex(\mathsf{S}_r) = d_r^+ - d_r^- \neq 0$) provided the system deficiency indices are equal ($\mathbf{d}^+ = \mathbf{d}^-$), as is illustrated in Example 8.2 that now follows.

Example 8.2. Consider the Lagrange symmetric differential expression

$$M_r[y] = i^3 y^{(3)}$$

on the semi-infinite interval $I_r = [0, \infty)$, for $r \in \Omega$, with the weight function $w(x) = 1$ for all $x \in [0, \infty)$. That is, for each $r \in \Omega$ (where $\text{card}(\Omega) = \aleph_0$) we use the same expression $M = i^3 y^{(3)}$ on $I = [0, \infty)$. Then a computation shows that the deficiency indices of the minimal operator generated by M_r in $L_r^2[0, \infty) := L^2[0, \infty)$ are

$$1 = d_r^+ \neq d_r^- = 2 \quad \text{for all} \quad r \in \Omega; \tag{8.46}$$

(compute the solutions of the differential equations $i^3 y^{(3)}(x) = \pm i y(x)$ with $x \in [0, \infty)$ and determine the number of linearly independent solutions in $L^2[0, \infty)$). Namely, for all $r \in \Omega$, we select a unit vector $e_+^{1,r}$ in the Hilbert subspace N_r^+; also linearly independent unit vectors $e_-^{1,r}, e_-^{2,r}$ for N_r^- (we surpress the subscript D_r in $[\cdot : \cdot]_{D_r}$ and in the D_r-norm, see Remark 6.1 and (6.20)).

For $e_+^{1,r}$ in N_r^+ we seek a function $e_+^{1,r}(x) = c \exp(-x)$ for all $x \in [0, \infty)$, where $c \in \mathbb{C}$, satisfying the requirement, for all $r \in \Omega$,

$$[e_+^{1,r} : e_+^{1,r}] = \int_0^\infty \left\{ \left(ie_+^{1,r}\right) \overline{e_+^{1,r}} - e_+^{1,r} \overline{\left(ie_+^{1,r}\right)} \right\} = 2i \int_0^\infty \left| e_+^{1,r} \right|^2 = i.$$

Then we compute

$$e_+^{1,r}(x) = \exp(-x) =: u_+^1(x) \text{ (say) for all } x \in [0, \infty). \tag{8.47}$$

For N_r^- we seek two basis vectors $e_-^{1,r}$ and $e_-^{2,r}$, each of the form, where $c_1, c_2 \in \mathbb{C}$,

$$c_1 \exp(-x/2) \cos\left(\sqrt{3}x/2\right) + c_2 \exp(-x/2) \sin\left(\sqrt{3}x/2\right);$$

satisfying, for all $r \in \Omega$ and $s = 1, 2$,

$$[e_-^{s,r} : e_-^{s,r}] = \int_0^\infty \left\{ \left(-ie_-^{s,r}\right) \overline{e_-^{s,r}} - e_-^{s,r} \overline{\left(-ie_-^{s,r}\right)} \right\} = -2i \int_0^\infty \left| e_-^{s,r} \right|^2 = -i.$$

It is immediate that we can select, for all $r \in \Omega$,

$$e_-^{1,r}(x) = \frac{2}{\sqrt{5}} \exp(-x/2) \cos\left(\sqrt{3}x/2\right) =: u_-^1(x) \text{ for all } x \in [0, \infty). \tag{8.48}$$

However the choice of, for all $x \in [0, \infty)$ and $r \in \Omega$,

$$E_-^{2,r}(x) = \frac{2}{\sqrt{3}} \exp(-x/2) \sin\left(\sqrt{3}x/2\right)$$

satisfies $[E_-^{2,r} : E_-^{2,r}] = -i$ but not the orthogonality condition since $[e_-^{1,r} : E_-^{2,r}] \neq 0$. Nevertheless we can select the vector $e_-^{2,r}$ by the Gram-Schmidt algorithm in order that $[e_-^{1,r} : e_-^{2,r}] = 0$ for all $r \in \Omega$, and then

$$e_-^{2,r}(x) = c\left(E_-^{2,r}(x) - e_-^{1,r}(x)/\sqrt{5}\right) =: u_-^2(x) \text{ for all } x \in [0, \infty). \tag{8.49}$$

Here the number c is determined to be $\sqrt{5}/2$ by the demand that $[e_-^{2,r} : e_-^{2,r}] = -i$ for all $r \in \Omega$.

Thus straightforward computations, following Theorem 6.1, yield

$$[e_+^{1,r} : e_+^{1,r}] = i \quad \text{so} \quad \left\langle e_+^{1,r}, e_+^{1,r} \right\rangle_+ = 1$$

$$[e_-^{1,r} : e_-^{1,r}] = -i\left\langle e_-^{1,r}, e_-^{1,r} \right\rangle_- = -i \quad \text{so} \quad \left\langle e_-^{1,r}, e_-^{1,r} \right\rangle_- = 1$$

$$[e_-^{2,r} : e_-^{2,r}] = -i\left\langle e_-^{2,r}, e_-^{2,r} \right\rangle_- = -i \quad \text{so} \quad \left\langle e_-^{2,r}, e_-^{2,r} \right\rangle_- = 1$$

and

$$[e_-^{1,r} : e_-^{2,r}] = \left\langle e_-^{1,r}, e_-^{2,r} \right\rangle_- = 0,$$

where all these results hold for all $r \in \Omega$.

Notwithstanding (8.46) the multi-interval system $\{I_r, M_r, w_r : r \in \Omega\}$ has the deficiency indices

$$\mathbf{d}^+ = \sum_{r \in \Omega} d_r^+ = \text{card}(\Omega) \qquad \mathbf{d}^- = \sum_{r \in \Omega} d_r^- = \text{card}(\Omega)$$

so that $\mathbf{d}^+ = \mathbf{d}^- = \mathbf{d}$ and the system symplectic excess $Ex = 0$, see [**11**]. Therefore the system minimal operator $\mathbf{T}_0$ on $\mathbf{D}(\mathbf{T}_0)$ in the system Hilbert space $\mathbf{H} = \sum_{r \in \Omega} \oplus L_r^2[0, \infty)$ does have self-adjoint extensions $\mathbf{T}$ on $\mathbf{D}(\mathbf{T}) \subset \mathbf{H}$, which correspond to complete Lagrangians L in $\mathsf{S} = \mathbf{D}(\mathbf{T}_1)/\mathbf{D}(\mathbf{T}_0)$, as in Section 6.

As an orthonormal basis $\{\mathbf{e}_+^t : t \in \mathbb{N}\}$ for $\mathbf{N}^+$ define, using (8.47), for all $t \in \mathbb{N}$

$$\mathbf{e}_+^t := \{0, 0, \ldots, 0, \overset{t\text{-th place}}{e_+^{1,t}}, 0, \ldots\} = \{0, 0, \ldots, 0, \overset{t\text{-th place}}{u_+^1}, 0, \ldots\}.$$

As an orthonormal basis $\{\mathbf{e}_-^t : t \in \mathbb{N}\}$ for $\mathbf{N}^-$ define, using (8.48) and (8.49), for all $t \in \mathbb{N}$ (and for all $p \in \mathbb{N}$)

$$\mathbf{e}_-^t := \{0, 0, \ldots, 0, \overset{t\text{-th place}}{e_-^{1,p}}, 0, \ldots\} = \{0, 0, \ldots, 0, \overset{t\text{-th place}}{u_-^1}, 0, \ldots\} \text{ for } t = 2p - 1$$

$$\mathbf{e}_-^t := \{0, 0, \ldots, 0, \overset{t\text{-th place}}{e_-^{2,p}}, 0, \ldots\} = \{0, 0, \ldots, 0, \overset{t\text{-th place}}{u_-^2}, 0, \ldots\} \text{ for } t = 2p.$$

We now determine a bijective correspondence, say U, between these orthonormal bases of $\mathbf{N}^-$ and $\mathbf{N}^+$ by defining $U\mathbf{e}_-^t := \mathbf{e}_+^t$ for all $t \in \mathbb{N}$, and then extend U to a unitary surjection $\mathbf{U}$ of $\mathbf{N}^-$ onto $\mathbf{N}^+$. As from (8.45), $\mathbf{U}$ now determines a complete Lagrangian $\mathsf{L} \subset \mathsf{S}$, and hence a self-adjoint extension $\mathbf{T}$ of $\mathbf{T}_0$ for the multi-interval system $\{I_r, M_r, w_r : r \in \mathbb{N}\}$ as defined above.

This example exhibits the phenomenon of "shearing"; to overcome the difference between the individual interval deficiency indices, see (8.46), the Lagrangian L has to associate the basis elements $e_+^{1,2p-1}, e_+^{1,2p}$ from the intervals I_{2p-1}, I_{2p} with the two basis elements $e_-^{1,p}$ and $e_-^{2,p}$ from the single interval I_p, and this holds for all $p \in \mathbb{N}$.

It is straightforward to modify this Example 8.2, by elementary transformations of the independent variable $x \in [0, \infty)$, so that the interval I_p is replaced by the interval $[p-1, p)$ for each $p \in \mathbb{N}$. In this sense all these intervals $[p-1, p)$ have a union which fills a single half-line $[0, \infty)$.

Example 8.3. Consider the Lagrange symmetric differential expression

$$M_r[y] = \operatorname{sgn}(r)\,(iy')$$

on the semi-infinite interval $I_r = [0, \infty)$, with given weight function $w_r(x) = 1$ for all $x \in [0, \infty)$. Here we take the index set $\Omega := \{r \in \mathbb{R} : r \neq 0\}$, so that $\operatorname{card}(\Omega) = \mathfrak{c} = \aleph_1$, on using the continuum hypothesis. Then a computation shows that the deficiency indices of the minimal operator generated by M_r in $L_r^2[0, \infty) := L^2[0, \infty)$ are

$$\text{(8.50)} \qquad \begin{cases} d_r^+ = 0 \quad \text{for all } r > 0 & d_r^- = 1 \quad \text{for all } r > 0 \\ = 1 \quad \text{for all } r < 0 & = 0 \quad \text{for all } r < 0. \end{cases}$$

Thus, from (8.50), whilst the minimal operator generated by each M_r in $L^2[0, \infty)$, for all $r \in \Omega$, fails to have equal deficiency indices, the multi-interval system $\{I_r, M_r, w_r : r \in \Omega\}$ has the system deficiency indices

$$\mathbf{d}^+ = \sum_{r \in \Omega} d_r^+ = \sum_{r<0} d_r^+ = \aleph_1$$

and

$$\mathbf{d}^- = \sum_{r \in \Omega} d_r^- = \sum_{r>0} d_r^- = \aleph_1$$

which are equal $\mathbf{d} = \mathbf{d}^\pm = \aleph_1$. Thus the multi-interval system does admit self-adjoint extensions $\mathbf{T}$ on $\mathbf{D}(\mathbf{T})$ of the system minimal operator $\mathbf{T}_0$ on $\mathbf{D}(\mathbf{T}_0) \subset \mathbf{H} = \sum_{r \in \Omega} \oplus L_r^2[0, \infty)$.

The same equality of the system deficiency indices holds if we restrict the index set to

$$\Omega = \{r \in [-1, 0)\} \cup \{r \in (0, 2]\},$$

but not if we take

$$\Omega = \{r \in [-1, 0)\} \cup \{r \in \mathbb{N}\}.$$

In each of these situations we can picture the differential expression $M_r = \operatorname{sgn}(r)\,(id/dx)$ as acting on certain functions $f : [0, \infty) \times \Omega \to \mathbb{C}$, where $[0, \infty) \times \Omega \subset \mathbb{R}^2$.

Example 8.4. This example concerns boundary complex symplectic spaces for finite multi-interval classical differential systems.

To simplify the details only one differential expression M is involved, *i.e.* for any interval $I \subseteq \mathbb{R}$

$$\text{(8.51)} \qquad M[f] := -f'' \text{ on } I;$$

the weight w is given by $w(x) = 1$ for all $x \in I$. The domain $D(M)$ of M is defined as

$$D(M) := \{f : I \to \mathbb{C} : f, f' \in AC_{\text{loc}}(I)\},$$

with the Green's formula, for all $[\alpha, \beta] \subseteq I$ and all $f, g \in D(M)$,

$$\int_\alpha^\beta \{M[f]\overline{g} - f\overline{M[g]}\} = [f, g](\beta) - [f, g](\alpha)$$

where, for M given by (8.51), $[\cdot, \cdot](\cdot) : I \times D(M) \times D(M) \to \mathbb{C}$ is determined by

$$[f, g](x) = f(x)\overline{g}'(x) - f'(x)\overline{g}(x) \text{ for all } x \in I. \tag{8.52}$$

For this Example 8.4 we define self-adjoint operators using the GKN-theory, see in particular [**14**], [**15**] and [**8**], and then use the results of Theorems 6.2 and 6.5 to construct the corresponding complete Lagrangian subspaces. However note that in calling on the results from [**14**], [**15**] and [**8**] there are minor changes in notations; for consistency we maintain the notations used in this paper.

We consider two boundary value problems:

1. Let $I = [-\pi, \pi]$ and the self-adjoint problem

$$M[y] \equiv -y''(x) = \lambda y(x) \text{ for all } x \in [-\pi, \pi] \tag{8.53}$$

$$y(-\pi) = y(\pi) = 0. \tag{8.54}$$

For the corresponding system $\{[-\pi, \pi], M, w \equiv 1\}$ the maximal domain $D(T_1) \subset L^2(-\pi, \pi)$ is defined by

$$D(T_1) := \{f : [-\pi, \pi] \to \mathbb{C} : f \in D(M) \text{ and } f, M[f] \in L^2(-\pi, \pi)\}, \tag{8.55}$$

with the operator T_1 defined by $T_1 f := M[f]$ for all $f \in D(T_1)$. We note that for this system $d^- = d^+ = 2$ so that there are self-adjoint extensions of the minimal operator $T_0 \subset T_1$, where

$$D(T_0) = \{f \in D(T_1) : f(-\pi) = f(\pi) = f'(-\pi) = f'(\pi) = 0\};$$

and all such extensions are determined by two boundary restrictions on the domain of the maximal operator T_1.

Note that for this first problem we here use the notation, see (2.19),

$$[f : g] := [f, g](\pi) - [f, g](-\pi) \tag{8.56}$$

for all $f, g \in D(T_1)$.

The domain of the self-adjoint operator T generated by the problem (8.53) and (8.54) is given by

$$D(T) = \{f \in D(T_1) : f(-\pi) = f(\pi) = 0\}. \tag{8.57}$$

In order to represent this domain $D(T)$ in terms of GKN-theory, see Theorem 6.5, and in terms of complete Lagrangian subspaces, see Theorem 6.2, we choose the boundary functions $\varphi, \psi \in D(T_1)$ as follows:

$$\varphi(x) = -(x + \pi) \text{ for all } x \in [-\pi, -3\pi/4] \text{ and } \operatorname{supp}(\varphi) \subset [-\pi, -\pi/2)$$
$$\psi(x) = (x - \pi) \text{ for all } x \in [3\pi/4, \pi] \text{ and } \operatorname{supp}(\psi) \subset (\pi/2, \pi].$$

For $f \in D(T_1)$ we have, recall (8.56),

$$[f : \varphi] \equiv -[f, \varphi](-\pi) = f(-\pi) \tag{8.58}$$

$$[f : \psi] \equiv [f, \psi](\pi) = f(\pi). \tag{8.59}$$

Clearly the pair $\{\varphi,\psi\}$ is linearly independent in $D(T_1)$ modulo $D(T_0)$ and the symmetry conditions hold, *i.e.* (see simplified notation in (8.56))

$$[\varphi:\varphi]=[\varphi:\psi]=[\psi:\varphi]=[\psi:\psi]=0.$$

Thus, from the GKN-theory, the domain $D(T)$ is given by

$$D(T)=\{f\in D(T_1):[f:\varphi]=[f:\psi]=0\} \tag{8.60}$$

with

$$Tf:=M[f] \text{ for all } f\in D(T). \tag{8.61}$$

Likewise the associated complete Lagrangian subspace $\mathsf{L}\subset\mathsf{S}$, with

$$\mathsf{S}=D(T_1)/D(T_0) \text{ and } \mathsf{L}=D(T)/D(T_0),$$

is determined by

$$\mathsf{L}=\{\mathsf{f}\in\mathsf{S}:[\mathsf{f}:\varphi]_{\mathsf{S}}=[\mathsf{f}:\psi]_{\mathsf{S}}=0\}, \tag{8.62}$$

where here φ and ψ now denote the corresponding cosets in S.

For this example we can give explicit details of the eigenvalues and eigenfunctions of the boundary value problem (8.53) and (8.54), equivalently the spectrum and eigenvectors of the operator T in $L^2(-\pi,\pi)$ as defined by (8.60) and (8.61).

Let $y:[-\pi,\pi]\times\mathbb{C}\to\mathbb{C}$ be any solution of the equation (8.53); then for some $A,B\in\mathbb{C}$

$$y(x,\lambda)=A\cos(x\sqrt{\lambda})+B\sin(x\sqrt{\lambda})/\sqrt{\lambda}$$

where, say, $0\le\arg(\sqrt{\lambda})<\pi$ when $0\le\arg(\lambda)<2\pi$. The application of the boundary conditions in the form (8.58) and (8.59) gives the following linear homogeneous equations to determine the numbers A and B, and the eigenvalues

$$A\cos(\pi\sqrt{\lambda})-B\sin(\pi\sqrt{\lambda})/\sqrt{\lambda}=0$$
$$A\cos(\pi\sqrt{\lambda})+B\sin(\pi\sqrt{\lambda})/\sqrt{\lambda}=0.$$

Thus the eigenvalues and eigenfunctions are, the latter for $x\in[-\pi,\pi]$,

$$\lambda_n^c=\left(n+\tfrac{1}{2}\right)^2 \qquad \psi_n^c(x)=\cos((n+1/2)\,x) \quad \text{for all } n\in\mathbb{N}_0$$
$$\lambda_n^s=n^2 \qquad \psi_n^s(x)=\sin(nx) \quad \text{for all } n\in\mathbb{N}.$$

We pass now to the second boundary value problem; this is a finite multi-interval system with $\Omega=\{1,2\}$;

2. Let $I_1=[-\pi,0]$ and $I_2=[0,\pi]$ with M_1 and M_2 both determined by (8.51) on their respective intervals.

The Definitions 5.1 to 5.4 give the notations for the operator theory in the system Hilbert function space, with $L_r^2=L^2(I_r)$ for $r=1,2$,

$$\mathbf{H}:=L_1^2\oplus L_2^2;$$

the maximal operator $\mathbf{T}_1$ has for its domain, following the notation of 1 above,

$$\mathbf{D}(\mathbf{T}_1):=\{\mathbf{f}=\{f_1,f_2\}\in\mathbf{H}: \begin{array}{l}(i)\ f_r\in D(T_{1,r}) \text{ for } r=1,2\\ (ii)\ \{T_{1,r}f_r:r=1,2\}\in\mathbf{H}\}\end{array} \tag{8.63}$$

and

$$\mathbf{T}_1\mathbf{f}:=\{T_{1,1}f_1,T_{1,2}f_2\} \text{ for all } \mathbf{f}=\{f_1,f_2\}\in\mathbf{D}(\mathbf{T}_1)\subset\mathbf{H}. \tag{8.64}$$

Here the minimal operator $\mathbf{T}_0$ on $\mathbf{D}(\mathbf{T}_0)\subseteq\mathbf{D}(\mathbf{T}_1)$ is given as usual.

The deficiency indices for this system are given by $\mathbf{d}^{\pm} = d_1^{\pm} + d_2^{\pm}$ so that the common index $\mathbf{d} = \mathbf{d}^- + \mathbf{d}^+ = 4$, since the differential expression M is regular on both I_1 and I_2. Thus the minimal operator $\mathbf{T}_0$ for this system has self-adjoint extensions determined by placing four boundary conditions on the maximal domain $\mathbf{D}(\mathbf{T}_1)$. We choose these four boundary conditions to give the following boundary value problem

$$M_1[y_1] = \lambda y_1 \text{ on } [-\pi, 0] \qquad M_2[y_2] = \lambda y_2 \text{ on } [0, \pi] \tag{8.65}$$

$$y_1(-\pi) = 0 = y_2(\pi) \tag{8.66}$$

$$y_1(0) = y_2(0) \qquad y_1'(0) = y_2'(0). \tag{8.67}$$

The operator restriction $\mathbf{T}$ of the maximal operator $\mathbf{T}_1$ determined by the boundary conditions (8.66) and (8.67) is given by

$$\begin{aligned}\mathbf{D}(\mathbf{T}) := \{\mathbf{f} \in \mathbf{D}(\mathbf{T}_1) : \ & (i)\ f_1(-\pi) = 0 = f_2(\pi) \\ & (ii)\ f_1(0) = f_2(0) \text{ and } f_1'(0) = f_2'(0)\}.\end{aligned} \tag{8.68}$$

In order to represent this domain in terms of the generalized GKN-theory, see Theorem 6.5, we introduce the four boundary condition vectors $\boldsymbol{\varphi}, \boldsymbol{\psi}, \boldsymbol{\kappa}, \boldsymbol{\theta}$ where $\boldsymbol{\varphi} = \{\varphi_1, \varphi_2\}$ *etc.*

To define the vectors $\boldsymbol{\varphi}$ and $\boldsymbol{\psi}$ we use the boundary condition functions φ and ψ as given in 1 above; let

$$\begin{cases} \varphi_1(x) := \varphi(x) & \text{for all } x \in [-\pi, 0] \\ \varphi_2(x) := 0 & \text{for all } x \in [0, \pi] \\ \psi_1(x) := 0 & \text{for all } x \in [-\pi, 0] \\ \psi_2(x) := \psi(x) & \text{for all } x \in [0, \pi]. \end{cases} \tag{8.69}$$

To define the vectors $\boldsymbol{\theta}$ and $\boldsymbol{\kappa}$ we introduce the functions θ and κ as follows:

$$\theta : [-\pi, \pi] \to \mathbb{R} \text{ with } \theta(x) = x \text{ for } x \in [-\pi/4, \pi/4] \text{ and } \operatorname{supp}(\theta) \subset [-\pi/2, +\pi/2]$$

$$\kappa : [-\pi, \pi] \to \mathbb{R} \text{ with } \kappa(x) = 1 \text{ for } x \in [-\pi/4, \pi/4] \text{ and } \operatorname{supp}(\kappa) \subset [-\pi/2, +\pi/2].$$

Now define

$$\begin{cases} \theta_1(x) := \theta(x) & \text{for all } x \in [-\pi, 0] \\ \theta_2(x) := \theta(x) & \text{for all } x \in [0, \pi] \\ \kappa_1(x) := \kappa(x) & \text{for all } x \in [-\pi, 0] \\ \kappa_2(x) := \kappa(x) & \text{for all } x \in [0, \pi]. \end{cases} \tag{8.70}$$

Recall that for vectors $\mathbf{f}, \mathbf{g} \in \mathbf{D}(\mathbf{T}_1)$

$$[\mathbf{f} : \mathbf{g}] = [f_1 : g_1]_1 + [f_2 : g_2]_2$$

where now for this two interval case, recall (8.52),

$$[f_1 : g_1]_1 = [f_1, g_1](0) - [f_1, g_1](-\pi) \text{ and } [f_2 : g_2]_2 = [f_2, g_2](\pi) - [f_2, g_2](0).$$

A calculation now shows the following results hold for the set of vectors $\{\boldsymbol{\varphi}, \boldsymbol{\psi}, \boldsymbol{\kappa}, \boldsymbol{\theta}\}$:

(1) The set is linearly independent in $\mathbf{D}(\mathbf{T}_1)$ modulo $\mathbf{D}(\mathbf{T}_0)$; see the definition (8.63).

(2) The symmetry conditions, using the simplied notation of (8.56),

$$[\boldsymbol{\varphi}:\boldsymbol{\varphi}] = [\boldsymbol{\psi}:\boldsymbol{\psi}] = [\boldsymbol{\kappa}:\boldsymbol{\kappa}] = [\boldsymbol{\theta}:\boldsymbol{\theta}] = 0$$

$$[\boldsymbol{\varphi}:\boldsymbol{\psi}] = [\boldsymbol{\psi}:\boldsymbol{\kappa}] = [\boldsymbol{\kappa}:\boldsymbol{\theta}] = [\boldsymbol{\theta}:\boldsymbol{\varphi}] = 0$$

$$[\boldsymbol{\varphi}:\boldsymbol{\kappa}] = [\boldsymbol{\psi}:\boldsymbol{\theta}] = 0$$

(3) For all $\mathbf{f} \in \mathbf{D}(\mathbf{T}_1)$

$$\begin{array}{lll} [\mathbf{f}:\boldsymbol{\varphi}] = 0 & \text{implies} & f_1(-\pi) = 0 \\ [\mathbf{f}:\boldsymbol{\psi}] = 0 & \text{implies} & f_2(-\pi) = 0 \\ [\mathbf{f}:\boldsymbol{\kappa}] = 0 & \text{implies} & f_1(0) = f_2(0) \\ [\mathbf{f}:\boldsymbol{\theta}] = 0 & \text{implies} & f_1'(0) = f_2'(0). \end{array}$$

Thus the set of vectors $\{\boldsymbol{\varphi}, \boldsymbol{\psi}, \boldsymbol{\kappa}, \boldsymbol{\theta}\}$ forms a maximal GKN-set for the boundary value problem (8.65), (8.66) and (8.67). Hence the operator $\mathbf{T}$ with domain $\mathbf{D}(\mathbf{T})$ defined by (8.68) and, where the operator $\mathbf{T}_1$ is defined by (8.64),

$$\mathbf{Tf} := \mathbf{T}_1\mathbf{f} \text{ for all } \mathbf{f} \in \mathbf{D}(\mathbf{T})$$

is self-adjoint in the vector Hilbert space $\mathbf{H} = L^2(-\pi, 0) \oplus L^2(0, \pi)$.

The associated complete Lagrangian subspace $\mathsf{L} \subset \mathsf{S}$, with

$$\mathsf{S} = \mathbf{D}(\mathbf{T}_1)/\mathbf{D}(\mathbf{T}_0) \text{ and } \mathsf{L} = \mathbf{D}(\mathbf{T})/\mathbf{D}(\mathbf{T}_0),$$

is determined by

$$\mathsf{L} = \{\mathsf{f} \in \mathsf{S} : [\mathsf{f}:\boldsymbol{\varphi}]_\mathsf{S} = [\mathsf{f}:\boldsymbol{\psi}]_\mathsf{S} = [\mathsf{f}:\boldsymbol{\kappa}]_\mathsf{S} = [\mathsf{f}:\boldsymbol{\theta}]_\mathsf{S} = 0\}.$$

For this example also we can give explicit details of the eigenvalues and eigenvectors of this boundary value problem.

Let $\mathbf{y} = \{y_1, y_2\}$ be a solution of the boundary value problem (8.65), (8.66) and (8.67); then $y_r : I_r \times \mathbb{C} \to \mathbb{C}$ and satisfy $M[y_r] = \lambda y_r$ on I_r for $r = 1, 2$, where λ is an eigenvalue. Thus we can write

$$y_1(x,\lambda) = A\cos(x\sqrt{\lambda}) + B\sin(x\sqrt{\lambda})/\sqrt{\lambda}$$
$$y_2(x,\lambda) = C\cos(x\sqrt{\lambda}) + D\sin(x\sqrt{\lambda})/\sqrt{\lambda}$$

where the numbers $A, B, C, D \in \mathbb{C}$ have to be determined. An application of the boundary conditions (8.66) yields the two linear equations

$$A\cos(\pi\sqrt{\lambda}) - B\sin(\pi\sqrt{\lambda})/\sqrt{\lambda} = 0$$
$$C\cos(\pi\sqrt{\lambda}) + D\sin(\pi\sqrt{\lambda})/\sqrt{\lambda} = 0.$$

The boundary conditions (8.67) require that

$$A = C \text{ and } B = D.$$

Taken together these four equations imply

$$A\cos(\pi\sqrt{\lambda}) - B\sin(\pi\sqrt{\lambda})/\sqrt{\lambda} = 0$$
$$A\cos(\pi\sqrt{\lambda}) + B\sin(\pi\sqrt{\lambda})/\sqrt{\lambda} = 0.$$

as for example 1 above; thus the eigenvalues for this problem 2 are identical with the eigenvalues for the first boundary value problem

Hence we have for the eigenvalues and eigenvectors of this second boundary value problem

$$\lambda_n^c = \left(n+\tfrac{1}{2}\right)^2 \qquad \mathbf{\Psi}_n^c = \{\cos((n+1/2)\,x), \cos((n+1/2)\,x)\} \text{ for all } n \in \mathbb{N}_0$$
$$\lambda_n^s = n^2 \qquad \mathbf{\Psi}_n^s = \{\sin(nx), \sin(nx)\} \text{ for all } n \in \mathbb{N}.$$

Remark 8.4.

(1) In retrospect it is not surprising that there is such a close connection between examples 1 and 2 of Example 8.4. The boundary conditions at the two end-points $-\pi$ and π are identical for both examples. In example 2 the interface conditions at $0\pm$ are chosen so the elements of the maximal domain $\mathbf{D}(\mathbf{T}_1)$, and their first derivatives, are continuous at the common point 0 of the two intervals I_1 and I_2.

(2) Note that whilst the operator $\mathbf{T}$ on the domain $\mathbf{D}(\mathbf{T}) \subset \mathbf{H}$ is self-adjoint in $\mathbf{H}$, neither of the restrictions of $\mathbf{T}$ to the constituent spaces $L^2(I_1)$ and $L^2(I_2)$ is self-adjoint in these spaces.

Bibliography

[1] N.I. Akhiezer and I.M. Glazman. *Theory of linear operators in Hilbert space*: Volumes **I** and **II** (Pitman and Scottish Academic Press, London and Edinburgh: 1981).

[2] H. Behncke and H. Focke. 'Deficiency indices of singular Schrödinger operators.' *Math. Z.* **158** (1978), 87-98.

[3] J.P. Boyd. 'Sturm-Liouville eigenvalue problems with an interior pole.' *J. Math. Physics* **22** (1981), 1575-1590.

[4] N. Dunford and J.T. Schwartz. *Linear operators:* Part **II** (Wiley, New York: 1963).

[5] W.N. Everitt. 'Linear ordinary quasi-differential expressions.' Proceedings of the 1983 Beijing Symposium on Differential Equations and Differential Geometry, 1–28. (Science Press, Beijing, P. R. China, 1986.)

[6] W.N. Everitt and L. Markus. 'Controllability of r-matrix quasi-differential equations.' *J. Differential Equations.* **89** (1991), 95-109.

[7] W.N. Everitt and L. Markus. 'Nonlinear quasi-differential control systems.' *Results Math.* **21** (1992), 65-82.

[8] W.N. Everitt and L. Markus. 'The Glazman-Krein-Naimark theorem for ordinary differential operators.' *New results in operator theory and its applications*, 118-130, *Oper. Theory Adv. Appl.*, **98** (Birkhäuser, Basel: 1997).

[9] W.N. Everitt and L. Markus. *Boundary value problems and symplectic algebra for ordinary differential and quasi-differential operators:* Mathematical Surveys and Monographs, **61**. (American Mathematical Society, Providence, RI: 1999).

[10] W.N. Everitt and L. Markus. 'Complex symplectic geometry with applications to ordinary differential operators.' *Trans. Amer. Math. Soc.* **351** (1999), 4905-4945.

[11] W.N. Everitt and L. Markus. 'Infinite dimensional complex symplectic spaces.' (To appear).

[12] W.N. Everitt and D. Race. 'Some remarks on linear ordinary quasidifferential expressions.' *Proc. London Math. Soc.* (3) **54** (1987), 300-320.

[13] W.N. Everitt, C. Shubin, G. Stolz and A. Zettl. 'Sturm-Liouville problems with an infinite number of interior singularities.' *Spectral theory and computational methods of Sturm-Liouville problems* (Knoxville, TN, 1996), 211–249. *Lecture Notes in Pure and Appl. Math.*, **191** (Dekker, New York: 1997).

[14] W.N. Everitt and A. Zettl. 'Sturm-Liouville differential operators in direct sum spaces.' *Rocky Mountain J. Math.* **16** (1986), 497-516.

[15] W.N. Everitt and A. Zettl. 'Differential operators generated by a countable number of quasi-differential expressions on the real line.' *Proc. London Math. Soc.* (3) **64** (1992), 524-544.

[16] F. Gesztesy. 'On the one-dimensional Coulomb Hamiltonian.' *J. Phys. A* **13** (1980), 867-875.

[17] F. Gesztesy and W. Kirsch. 'One-dimensional Schrödinger operators with interactions on a discrete set.' *J. Reine Angew. Math.* **362** (1985), 28-50.

[18] F. Gesztesy, C. Macedo and L. Streit. 'An exactly solvable periodic Schrödinger operator.' *J. Phys. A* **18** (1985), L503-L507.

[19] V. Guillemin and S. Sternberg. *Symplectic techniques in physics:* (Cambridge University Press:1990).

[20] M.S. Homer. 'Spectral properties of the Laplace tidal wave equation.' *J. London Math. Soc.* (2) **45** (1992).

[21] M.S. Homer. 'Boundary value problems for the Laplace tidal wave equation.' *Proc. Roy. Soc. London Ser. A* **428** (1990), 157-180.

[22] L. Markus. *Hamiltonian dynamics and symplectic manifolds:* Lecture Notes, University of Minnesota. (University of Minnesota Bookstores; 1973, 1-256).

[23] L. Markus. 'Control of quasi-differential equations.' *Ann. Polon. Math.* **51** (1990), 229-239.

[24] M.A. Naimark. *Linear differential operators:* Part **II** (Ungar, New York: 1968).
[25] N. Steenrod. *The topology of fibre bundles:* (Princeton University Press, Princeton: 1951).
[26] J. Weidmann. *Linear operators in Hilbert space:* (Springer-Verlag, Heidelberg: 1980).
[27] A. Zettl. 'Adjoint and selfadjoint boundary value problems with interface conditions.' *SIAM J. Appl. Math.* **16** (1968), 851-859.
[28] A. Zettl. 'Formally self-adjoint quasi-differential operators.' *Rocky Mountain J. Math.* **5** (1975), 453-474.

Editorial Information

To be published in the *Memoirs*, a paper must be correct, new, nontrivial, and significant. Further, it must be well written and of interest to a substantial number of mathematicians. Piecemeal results, such as an inconclusive step toward an unproved major theorem or a minor variation on a known result, are in general not acceptable for publication. Papers appearing in *Memoirs* are generally longer than those appearing in *Transactions*, which shares the same editorial committee.

As of January 31, 2001, the backlog for this journal was approximately 7 volumes. This estimate is the result of dividing the number of manuscripts for this journal in the Providence office that have not yet gone to the printer on the above date by the average number of monographs per volume over the previous twelve months, reduced by the number of volumes published in four months (the time necessary for preparing a volume for the printer). (There are 6 volumes per year, each containing at least 4 numbers.)

A Consent to Publish and Copyright Agreement is required before a paper will be published in the *Memoirs*. After a paper is accepted for publication, the Providence office will send a Consent to Publish and Copyright Agreement to all authors of the paper. By submitting a paper to the *Memoirs*, authors certify that the results have not been submitted to nor are they under consideration for publication by another journal, conference proceedings, or similar publication.

Information for Authors

Memoirs are printed from camera copy fully prepared by the author. This means that the finished book will look exactly like the copy submitted.

The paper must contain a *descriptive title* and an *abstract* that summarizes the article in language suitable for workers in the general field (algebra, analysis, etc.). The *descriptive title* should be short, but informative; useless or vague phrases such as "some remarks about" or "concerning" should be avoided. The *abstract* should be at least one complete sentence, and at most 300 words. Included with the footnotes to the paper should be the 2000 *Mathematics Subject Classification* representing the primary and secondary subjects of the article. The classifications are accessible from `www.ams.org/msc/`. The list of classifications is also available in print starting with the 1999 annual index of *Mathematical Reviews*. The Mathematics Subject Classification footnote may be followed by a list of *key words and phrases* describing the subject matter of the article and taken from it. Journal abbreviations used in bibliographies are listed in the latest *Mathematical Reviews* annual index. The series abbreviations are also accessible from `www.ams.org/publications/`. To help in preparing and verifying references, the AMS offers MR Lookup, a Reference Tool for Linking, at `www.ams.org/mrlookup/`. When the manuscript is submitted, authors should supply the editor with electronic addresses if available. These will be printed after the postal address at the end of the article.

Electronically prepared manuscripts. The AMS encourages electronically prepared manuscripts, with a strong preference for $\mathcal{AMS}$-LaTeX. To this end, the Society has prepared $\mathcal{AMS}$-LaTeX author packages for each AMS publication. Author packages include instructions for preparing electronic manuscripts, the *AMS Author Handbook*, samples, and a style file that generates the particular design specifications of that publication series. Though $\mathcal{AMS}$-LaTeX is the highly preferred format of TeX, author packages are also available in $\mathcal{AMS}$-TeX.

Authors may retrieve an author package from e-MATH starting from `www.ams.org/tex/` or via FTP to `ftp.ams.org` (login as `anonymous`, enter username as password, and type `cd pub/author-info`). The *AMS Author Handbook* and the *Instruction Manual* are available in PDF format following the author packages link from `www.ams.org/tex/`. The author package can be obtained free of charge by sending email to `pub@ams.org` (Internet) or from the Publication Division, American Mathematical Society, P.O. Box 6248, Providence, RI 02940-6248. When requesting an author package, please specify $\mathcal{A}\mathcal{M}\mathcal{S}$-LaTeX or $\mathcal{A}\mathcal{M}\mathcal{S}$-TeX, Macintosh or IBM (3.5) format, and the publication in which your paper will appear. Please be sure to include your complete mailing address.

Sending electronic files. After acceptance, the source file(s) should be sent to the Providence office (this includes any TeX source file, any graphics files, and the DVI or PostScript file).

Before sending the source file, be sure you have proofread your paper carefully. The files you send must be the EXACT files used to generate the proof copy that was accepted for publication. For all publications, authors are required to send a printed copy of their paper, which exactly matches the copy approved for publication, along with any graphics that will appear in the paper.

TeX files may be submitted by email, FTP, or on diskette. The DVI file(s) and PostScript files should be submitted only by FTP or on diskette unless they are encoded properly to submit through email. (DVI files are binary and PostScript files tend to be very large.)

Electronically prepared manuscripts can be sent via email to `pub-submit@ams.org` (Internet). The subject line of the message should include the publication code to identify it as a Memoir. TeX source files, DVI files, and PostScript files can be transferred over the Internet by FTP to the Internet node `e-math.ams.org` (130.44.1.100).

Electronic graphics. Comprehensive instructions on preparing graphics are available at `www.ams.org/jourhtml/graphics.html`. A few of the major requirements are given here.

Submit files for graphics as EPS (Encapsulated PostScript) files. This includes graphics originated via a graphics application as well as scanned photographs or other computer-generated images. If this is not possible, TIFF files are acceptable as long as they can be opened in Adobe Photoshop or Illustrator. No matter what method was used to produce the graphic, it is necessary to provide a paper copy to the AMS.

Authors using graphics packages for the creation of electronic art should also avoid the use of any lines thinner than 0.5 points in width. Many graphics packages allow the user to specify a "hairline" for a very thin line. Hairlines often look acceptable when proofed on a typical laser printer. However, when produced on a high-resolution laser imagesetter, hairlines become nearly invisible and will be lost entirely in the final printing process.

Screens should be set to values between 15% and 85%. Screens which fall outside of this range are too light or too dark to print correctly. Variations of screens within a graphic should be no less than 10%.

Inquiries. Any inquiries concerning a paper that has been accepted for publication should be sent directly to the Electronic Prepress Department, American Mathematical Society, P. O. Box 6248, Providence, RI 02940-6248.

Editors

Selected Titles in This Series

(*Continued from the front of this publication*)

685 **Jean-Luc Joly, Guy Metivier, and Jeffrey Rauch,** Caustics for dissipative semilinear oscillations, 2000

684 **Harvey I. Blau, Bangteng Xu, Z. Arad, E. Fisman, V. Miloslavsky, and M. Muzychuk,** Homogeneous integral table algebras of degree three: A trilogy, 2000

683 **Serge Bouc,** Non-additive exact functors and tensor induction for Mackey functors, 2000

682 **Martin Majewski,** ational homotopical models and uniqueness, 2000

681 **David P. Blecher, Paul S. Muhly, and Vern I. Paulsen,** Categories of operator modules (Morita equivalence and projective modules, 2000

680 **Joachim Zacharias,** Continuous tensor products and Arveson's spectral C^*-algebras, 2000

679 **Y. A. Abramovich and A. K. Kitover,** Inverses of disjointness preserving operators, 2000

678 **Wilhelm Stannat,** The theory of generalized Dirichlet forms and its applications in analysis and stochastics, 1999

677 **Volodymyr V. Lyubashenko,** Squared Hopf algebras, 1999

676 **S. Strelitz,** Asymptotics for solutions of linear differential equations having turning points with applications, 1999

675 **Michael B. Marcus and Jay Rosen,** Renormalized self-intersection local times and Wick power chaos processes, 1999

674 **R. Lawther and D. M. Testerman,** A_1 subgroups of exceptional algebraic groups, 1999

673 **John Lott,** Diffeomorphisms and noncommutative analytic torsion, 1999

672 **Yael Karshon,** Periodic Hamiltonian flows on four dimensional manifolds, 1999

671 **Andrzej Rosłanowski and Saharon Shelah,** Norms on possibilities I: Forcing with trees and creatures, 1999

670 **Steve Jackson,** A computation of δ^1_5, 1999

669 **Seán Keel and James McKernan,** Rational curves on quasi-projective surfaces, 1999

668 **E. N. Dancer and P. Poláčik,** Realization of vector fields and dynamics of spatially homogeneous parabolic equations, 1999

667 **Ethan Akin,** Simplicial dynamical systems, 1999

666 **Mark Hovey and Neil P. Strickland,** Morava K-theories and localisation, 1999

665 **George Lawrence Ashline,** The defect relation of meromorphic maps on parabolic manifolds, 1999

664 **Xia Chen,** Limit theorems for functionals of ergodic Markov chains with general state space, 1999

663 **Ola Bratteli and Palle E. T. Jorgensen,** Iterated function systems and permutation representation of the Cuntz algebra, 1999

662 **B. H. Bowditch,** Treelike structures arising from continua and convergence groups, 1999

661 **J. P. C. Greenlees,** Rational S^1-equivariant stable homotopy theory, 1999

660 **Dale E. Alspach,** Tensor products and independent sums of $\mathcal{L}_p$-spaces, $1 < p < \infty$, 1999

659 **R. D. Nussbaum and S. M. Verduyn Lunel,** Generalizations of the Perron-Frobenius theorem for nonlinear maps, 1999

658 **Hasna Riahi,** Study of the critical points at infinity arising from the failure of the Palais-Smale condition for n-body type problems, 1999

657 **Richard F. Bass and Krzysztof Burdzy,** Cutting Brownian paths, 1999

656 **W. G. Bade, H. G. Dales, and Z. A. Lykova,** Algebraic and strong splittings of extensions of Banach algebras, 1999

For a complete list of titles in this series, visit the AMS Bookstore at **www.ams.org/bookstore/**.

[redacted]